Mammalogy

Mammalogy

Kate Porter

Larsen & Keller
www.larsen-keller.com

Mammalogy
Kate Porter
ISBN: 978-1-64172-511-8 (Hardback)

Larsen & Keller

Published by Larsen and Keller Education,
5 Penn Plaza,
19th Floor,
New York, NY 10001, USA

Cataloging-in-Publication Data

Mammalogy / Kate Porter.
     p. cm.
Includes bibliographical references and index.
ISBN 978-1-64172-511-8
1. Mammalogy. 2. Mammals. 3. Zoology. I. Porter, Kate.
QL703 .M36 2020
599--dc23

For more information regarding Larsen and Keller Education and its products, please visit the publisher's website www.larsen-keller.com

TABLE OF CONTENTS

This book has been written, keeping in view that students want more practical information. Thus, my aim has been to make it as comprehensive as possible for the readers. I would like to extend my thanks to my family and co-workers for their knowledge, support and encouragement all along.

Mammalogy is a sub-division of vertebrate zoology which includes the study of mammals. It is a class of vertebrates that have characteristics such as homeothermic metabolism, four-chambered hearts, fur and complex nervous system. There are various branches of mammalogy including taxonomy and systematics, natural history, ecology, ethology, anatomy and physiology, and management and control. It also branches into other sub-disciplines that are taxonomically-oriented such as cetology and primatology. The scientific study of primates is known as primatology. It includes the study of both living and extinct primates in their natural habitats as well as in laboratories. Cetology studies various species of whales, porpoises and dolphins in the scientific order Cetacea. This book provides comprehensive insights into the field of mammalogy. Some of the diverse topics covered herein address the varied branches that fall under this category. The topics covered in this book offer the readers new insights in this field.

A brief description of the chapters is provided below for further understanding:

Chapter – Introduction to Mammalogy

Mammals refer to the class of vertebrate animals which have fur and mammary glands. Mammalogy deals with the scientific study of mammals including their characteristics such as homeothermic metabolism, and complex nervous systems. This is an introductory chapter which will briefly introduce all the significant aspects of mammalogy.

Chapter – Sub-disciplines of Mammalogy

There are many sub-disciplines that fall under the domain of mammalogy which include primatology, cetology, hippology, cynology, felinology, etc. This chapter closely examines these sub-disciplines of mammalogy to provide an extensive understanding of the subject.

Chapter – Anatomy and Physiology of Mammals

Anatomy and physiology are the branches of science that deal with bodily structure and functioning of living organisms respectively. It includes the study of mammals such as cetacea, equine, felidae, etc. This chapter delves into these aspects associated with mammalogy for providing an in-depth understanding of the subject.

Chapter – The Behavior of Mammals

Behavior is the way in which a living organism acts or conducts itself due to the stimulus presented by its surrounding. Behavioral aspects of mammals such as cetacea, equine, felidae, etc. are thoroughly discussed in this chapter. This chapter has been carefully written to provide an easy understanding of the various concepts related to behavior of mammals.

Chapter – Diseases of Mammals

There are various diseases which are caused in different mammals like horses, dogs, cats, etc. Potomac horse fever, skin cancer, infectious canine hepatitis, hypertrophic osteodystrophy, congenital sensorineural deafness, feline asthma, etc. are some of these diseases. The topics elaborated in this chapter will help in gaining a better perspective about these diseases caused in mammals.

Kate Porter

Introduction to Mammalogy

- **Mammal**
- **Mammalogy**

Mammals refer to the class of vertebrate animals which have fur and mammary glands. Mammalogy deals with the scientific study of mammals including their characteristics such as homeothermic metabolism, and complex nervous systems. This is an introductory chapter which will briefly introduce all the significant aspects of mammalogy.

Mammal

Mammals are the class of vertebrate animals characterized by the presence of hair (or fur) and mammary glands, which in females produce milk for the nourishment of young. The other extant (living) classes of vertebrates (animals with backbones) include fish (with a few recognized classes), amphibians, reptiles, and birds.

Like birds, mammals are endothermic or "warm-blooded," and have four-chambered hearts. Mammals also have a diaphragm, a muscle below the rib cage that aids breathing. Some other vertebrates have a diaphragm, but mammals are the only vertebrates with a prehepatic diaphragm, that is, in front of the liver. Mammals are also the only vertebrates with a single bone in the lower jaw.

The choice of the word mammal to describe the class—rather than the presence of hair or a single bone in the lower jaw—is an interesting choice. In the eighteenth century, some scientists did refer to this group as hairy things, or "Pilosa" (now a designation for a group of placental mammals, including anteaters and sloths). But Carolus Linneaus provided the term Mammalia. Some authorities trace Linneaus choice to his advocacy of mothers' breastfeeding their own children, and indeed Linneaus authored a book on the issue. Whatever the reason, this terminology ties mammals to a feature that is connected to a key mammalian characteristic: parental behavior. Although caring for young is evident in many animals, including crocodiles, it reaches a zenith in birds and mammals. Among the primates, human mothers spend more time with their infants than any other species.

There are three major subdivisions of mammals: monotremes, marsupials, and placentals. Monotremes are mammals that lay eggs, and include the platypuses and echidnas (spiny anteaters). Marsupials are generally characterized by the female having a pouch in which it rears its young through early infancy, as well as various reproductive traits that distinguish them from other mammals. Opossums, wombats, kangaroos, and wallabies are examples of marsupials. Placentals generally can be distinguished from other mammals in that the fetus is nourished during gestation via a placenta, although bandicoots (marsupial omnivores) are a conspicuous exception to this rule. About 5,500 living species of mammals have been identified. Phylogenetically (classification based on evolutionary relatedness), Class Mammalia is defined as all descendants of the most recent common ancestor of monotremes and the therian mammals (Theria is a taxon that includes the closely affiliated marsupials and placentals).

Characteristics

Although all mammals are endothermic, this is not a defining feature of mammals, since this trait is shared with birds. However, the presence of hair is a unique mammalian characteristic. This filamentous outgrowth of the skin projects from the epidermis, though it grows from follicles deep in the dermis. In non-human species, hair is commonly referred to as fur. The presence of hair has helped mammals to maintain a stable core body temperature. Hair and endothermy has aided mammals in inhabiting a wide diversity of environments, from deserts to polar environments, and be active daytime and nighttime.

The amount of hair reflects the environment to which the animal is adapted. Polar bears have thick, water-repellent fur with hollow hairs that trap heat well. Whales have very limited hair in isolated areas, thus reducing drag in the water. Instead, they maintain internal temperatures with a thick layer of blubber (vascularized fat).

No mammals have hair that is naturally blue or green in color. Some cetaceans (whales, dolphins and porpoises), along with the mandrills, appear to have shades of blue skin. Many mammals are indicated as having blue hair or fur, but in all cases it will be found to be a shade of gray. The two-toed sloth can seem to have green fur, but this color is caused by algal growths.

Although monotremes are endothermic, their metabolic rates and body temperature are lower than that of most other mammals. Monotremes maintain an average temperature of around 32°C (90°F) compared to about 35°C (95°F) for marsupials and 38°C (100°F) for most placentals.

Mammals have integumentary systems made up of three layers: the outermost epidermis, the dermis, and the hypodermis. This characteristic is not unique to mammals, but is found in all vertebrates. The epidermis is typically ten to thirty cells thick, its main function being to provide a waterproof layer. Its outermost cells are constantly lost; its bottommost cells are constantly dividing and pushing upward. The middle layer, the dermis, is fifteen to forty times thicker than the epidermis. The dermis is made up of many components, such as bony structures and blood vessels. The hypodermis is made up of adipose tissue. Its job is to store lipids and to provide cushioning and insulation. The thickness of this layer varies widely from species to species.

Along with hair, the presence of mammary glands, for feeding milk to their young, is another defining feature of mammals. The monotremes do not have nipples, but they do have mammary glands. The milk is secreted via the skin onto a surface, where it can be sucked or lapped up by the newborn.

Most mammals give birth to live young, but the monotremes lay eggs. Live birth also occurs in some non-mammalian species, such as guppies and hammerhead sharks; thus, it is not a distinguishing characteristic of mammals.

Mammals have three bones in each ear and one (the dentary) on each side of the lower jaw; all other vertebrates with ears have one bone (the stapes) in the ear and at least three on each side of the jaw. A group of therapsids called cynodonts had three bones in the jaw, but the main jaw joint was the dentary and the other bones conducted sound. The extra jawbones of other vertebrates are thought to be homologous with the malleus and incus of the mammal ear.

Northern right whale dolphin (*L. borealis*).

All mammalian brains possess a neocortex (or neopallium) that is involved in higher functions, such as sensory perception, generation of motor commands, spatial reasoning, and in humans, language and conscious thought. This brain region is unique to mammals (as is a single bone in the lower jaw, and the prehepatic diaphragm, mentioned above).

Most mammals are terrestrial, but some are aquatic, including sirenia (manatees and dugongs) and the cetaceans. Whales are the largest of all animals. There are semi-aquatic species, such as seals, which come to land to breed but spend most of the time in water. Most cetaceans live in salt water, but there are some dolphin species that live in fresh water, such as the Amazon River Dolphin (*Inia geoffrensis*) that lives in the Amazon and Orinoco River basins.

True flight has evolved only once in mammals, the bats; mammals such as flying squirrels and flying lemurs are actually gliding animals.

Habitat, Physiological Characteristics and Behavior

Different species of mammals have evolved to live in nearly all terrestrial and aquatic habitats on Earth. Mammals inhabit every terrestrial biome, from deserts to tropical rainforests to polar icecaps. Many species are arboreal, spending most or all of their time in the forest canopy. One group (bats) has even evolved powered flight, only the third time that this ability has evolved in vertebrates (the other two groups being birds and extinct Pterosaurs).

Many mammals are partially aquatic, living near lakes, streams, or the coastlines of oceans (e.g., seals, sea lions, walruses, otters, muskrats, and many others). Whales and dolphins (Cetacea) are fully aquatic, and can be found in all oceans of the world, and some rivers. Whales can be found in polar, temperate, and tropical waters, both near shore and in the open ocean, and from the water's surface to depths of over 1 kilometer.

All mammals have hair at some point during their development, and most have hair their entire lives. Adults of some species lose most or all of their hair but, even in mammals like whales and dolphins, hair is present at least during some phase of ontogeny. Mammalian hair, made of a protein called keratin, serves at least four functions. First, it slows the exchange of heat with the environment (insulation). Second, specialized hairs (whiskers or "vibrissae") have a sensory function, letting an animal know when it is in contact with an object in its environment. Vibrissae are often richly innervated and well supplied with muscles that control their position. Third, hair affects appearance through its color and pattern, helping to camouflage predators or prey as well as signal to predators a defensive mechanism (for example, the conspicuous color pattern of a skunk is a warning to predators). Hair also communicates social information (for example, threats, such as the erect hair on the back of a wolf; sex, such as the different colors of male and female capuchin monkeys; or the presence of danger, such as the white underside of the tail of a white-tailed deer). Fourth, hair provides some protection, simply by providing an additional protective layer (against abrasion or sunburn, for example) or by taking on the form of dangerous spines that deter predators (porcupines, spiny rats, others).

Mammals are typically characterized by their highly differentiated teeth. Teeth are replaced just once during a mammal's life (a condition called diphyodonty). Other characteristics found in most mammals include: a lower jaw made up of a single bone, the dentary; four-chambered hearts; a secondary palate separating air and food passages in the mouth; a muscular diaphragm separating thoracic and abdominal cavities; a highly developed brain; endothermy and homeothermy; separate sexes, with the sex of an embryo being determined by the presence of a Y or 2 X chromosomes; and internal fertilization.

Development

A Short-beaked echidna (Tachyglossus aculeatus), an egg laying mammal.

A Koala (Phascolarctos cinereus) on a tree.

A mother Capybara (Hydrochoerus hydrochaeris) nursing her litter.

There are three major groups of mammals, each united by a major feature of embryonic development. Monotremes (Prototheria) lay eggs, which is the most primitive reproductive condition in mammals. Marsupials (Metatheria) give birth to highly altricial young after a very short gestation period (8 to 43 days). The young are born at a relatively early stage of morphological development. They attach to the mother's nipple and spend a proportionally greater amount of time nursing as they develop. Gestation lasts much longer in placental mammals (Eutheria). During gestation, eutherian young interact with their mother through a placenta, a complex organ that connects the embryo with the uterus. Once born, all mammals are dependent upon their mothers for milk. Aside from these few generalities, mammals exhibit a diversity of developmental and life history patterns that vary among species and larger taxonomic groups.

Reproduction

Most mammalian species are either polygynous (one male mates with multiple females) or promiscuous (both males and females have multiple mates in a given reproductive season). Because females spend a long period of time in gestation and lactation, it is often the case that males can produce many more offspring in a mating season than can females. Thus, the most common mammalian mating system is polygyny, with relatively few males fertilizing multiple females and many males fertilizing none. This causes intense competition between males in many species, and the potential for females to be discriminating in choice of sire for offspring, leading to complex behaviors and morphologies associated with reproduction. Many mammal groups are marked by sexual dimorphism as a result of selection for males that can better compete for access to females.

About 3 percent of mammalian species are monogamous, males mating with a single female each season. In these cases, males provide at least some care to their offspring. Often, mating systems may vary within species depending upon local environmental conditions. For example, when resources are low, males may mate with a single female and provide care for the young. When resources are abundant, the mother may be able to care for young on her own and males will attempt to sire offspring with multiple females.

Other mating systems, such as polyandry, can be found among mammals. Some species (e.g. common marmosets and African lions) display cooperative breeding, in which groups of females, and sometimes males, share the care of young from one or more females. Naked mole rats have a unique mating system among mammals. Like social insects (Hymenoptera and Isoptera), naked mole rats are eusocial, with a queen female mating with several males and bearing all of the young in the colony. Other colony members assist in the care of her offspring and do not reproduce themselves.

Mammals range from many altricial young in each bout of reproduction (rodents and insectivores) to those species that give birth to one or a few precocial young. The former tend toward high early mortality and short lifespans, while the latter invest energy in a few offspring that develop into efficient competitors, living longer in generally stable environments (Cetaceans, primates and artiodactyls). Most mammals make use of a den or nest for the protection of their young. Some mammals, however, are born well developed and able to locomote on their own soon after birth. Most notable in this regard are artiodactyls such as wildebeest or giraffes. Cetacean young must also swim on their own shortly after birth.

Generally, smaller mammals live short lives and larger mammals live longer lives. Bats (Chiroptera) are an exception to this pattern, being relatively small mammals that can live for one or more decades in natural conditions, considerably longer than natural lifespans of significantly larger mammals. Mammalian lifespans range from one year or less to 70 or more years in the wild. Bowhead whales may live more than 200 years.

Behavior

There are mammal species that exhibit nearly every type of lifestyle, including fossorial, aquatic, terrestrial, and arboreal lifestyles. Locomotion styles are also diverse: mammals may swim, run, bound, fly, glide, burrow, or climb as a means of moving throughout their environment.

Social behavior and activity patterns vary considerably as well. Some mammals live in groups of tens, hundreds, thousands or more individuals. Other mammals are generally solitary except when mating or raising young. Mammals may be nocturnal, diurnal, or crepuscular.

Olfaction, hearing, tactile perception, and vision are all important sensory modalities in mammals. Olfaction plays a key role in many aspects of mammalian ecology, including foraging, mating and social communication. Mammalian hearing is well developed as well. In some species, it is the primary form of perception. Echolocation, the ability to perceive objects in the external environment by listening to echoes from sounds generated by an animal, has evolved in several groups, including microchiropteran bats (Chiroptera) and many toothed whales and dolphins (Odontoceti). Vision is well developed in a large number of mammals, although less important in many species that live underground or use echolocation.

Mammals can be carnivores (e.g., most species within Carnivora), herbivores (e.g., Perissodactyla, Artiodactyla), or omnivores (e.g., many primates). Mammals eat both invertebrates and vertebrates (including other mammals), plants (including fruit, nectar, foliage, wood, roots, and seeds) and fungi. Being endotherms, mammals require much more food than ectotherms of similar proportions. Thus, relatively few mammals can have a large impact on the populations of their food items.

Economic Impact on Humans

Mammals are a vital economic resource for humans. Many mammals have been domesticated to provide products such as meat and milk (e.g., cows and goats) or fiber (sheep and alpacas). Many mammals are kept as service animals or pets (e.g., dogs, cats, ferrets). Mammals are important for the ecotourism industry as well: people travel to zoos and throughout the world to see animals like elephants, lions, or whales. Bats often help control populations of crop pests. Norway rats and domestic mice are vitally important in medical and other scientific research, serving as models in human medicine and research.

On the other hand, some species have a detrimental impact on humans and the ecosystem. Many mammals that eat fruit, seeds, and other types of vegetation are crop pests. Carnivores are often a threat to livestock and even to human lives. Mammals common in urban or suburban areas have caused automobile accidents when straying into roads and have become household pests. A few species coexist exceptionally well with people, including some feral domesticated mammals (e.g., rats, house mice, pigs, cats, and dogs). As a result of intentional or unintentional introductions near human habitation, these animals have had considerable negative impacts on the local biota of many regions of the world, especially the endemic biota of oceanic islands.

Many mammals can transmit diseases to humans or livestock. The bubonic plague is perhaps the most well-known example, spread via fleas carried by rodents. Rabies, which can be transmitted among mammalian species, is also a significant threat to livestock and can kill humans as well.

Conservation Status

Overexploitation, habitat destruction and fragmentation, the introduction of exotic species, and other anthropogenic pressures threaten mammals worldwide. In the past five centuries, at least 82 mammal species have gone extinct. Most mammals have been assessed for conservation concern and currently, the International Union for Conservation of Nature and Natural Resources (IUCN) has listed about 1000 species (roughly 25% of all known mammals), as being at some risk of extinction. Species that are naturally rare or require large home ranges are often at risk due to habitat loss and fragmentation. Species that are seen to threaten humans, livestock, or crops may be directly targeted for extirpation. Those species that are exploited by humans as a resource (e.g., for their meat or fur) but are not domesticated are often depleted to critically low levels. Finally, global climate change is adversely affecting many mammals. The geographic ranges of many mammals are shifting as species lose the ability to adjust to increasingly rapid changes in local temperatures and climate.

Mammalogy

Mammalogy is scientific study of mammals. Interest in nonhuman mammals dates far back in prehistory, and the modern science of mammalogy has its broad foundation in the knowledge of mammals possessed by primitive peoples. The ancient Greeks were among the first peoples to write systematically on mammalian natural history, and they knew many mammals not native to Greece; Aristotle recognized that whales and dolphins (cetaceans), although fishlike in form, are mammals allied to terrestrial furbearers. Until the late 18th century, much scientific work on mammals was devoted to taxonomy or to the practical matters of animal husbandry. The scientific explorations of the 19th century resulted in large collections of specimens from virtually all parts of the world. Most of the world's mammal species are believed to be known to science (with the possible exception of a good many rodent and bat species), but the biology of many species is totally unknown. Modern mammalogy is a multidisciplinary field, encompassing specialists in anatomy, paleontology, ecology, behaviour, and many other areas.

Mammalian taxonomy traditionally relied largely on museum collections of preserved skins (with their skulls), but, by the second half of the 20th century, additional information was being gained from other studies—e.g., behaviour, genetics, and biochemistry. In both laboratory and field research, new techniques and instruments have opened avenues of research that had previously been difficult or impossible. The self-contained underwater breathing apparatus (scuba), for example, has been important in many aspects of marine mammalogy. Telemetry, the use of minute radio transmitters to convey information to the researcher from a free-living animal, has been a particularly useful tool, allowing the tracking of the animal in its natural state and the monitoring of physiological information. Video technology also has come into use, while in the laboratory a rapidly increasing array of molecular techniques have changed the way mammalogists determine evolutionary relationships (phylogeny) as well.

Sub-disciplines of Mammalogy

2

- **Primatology**
- **Cetology**
- **Hippology**
- **Cynology**
- **Felinology**

There are many sub-disciplines that fall under the domain of mammalogy which include primatology, cetology, hippology, cynology, felinology, etc. This chapter closely examines these sub-disciplines of mammalogy to provide an extensive understanding of the subject.

Primatology

Primatology is the study of the primate order of mammals—other than recent humans (Homo sapiens). The species are characterized especially by advanced development of binocular vision, specialization of the appendages for grasping, and enlargement of the cerebral hemispheres.

Nonhuman primates provide a broad comparative framework within which physical anthropologists can study aspects of the human career and condition. Comparative morphological studies, particularly those that are complemented by biomechanical analyses, provide major clues to the functional significance and evolution of the skeletal and muscular complexes that underpin humans' bipedalism, dextrous hands, bulbous heads, outstanding noses, and puny jaws. The wide variety of adaptations that primates have made to life in trees and on the ground are reflected in their limb proportions and relative development of muscles.

Free-ranging primates exhibit a trove of physical and behavioral adaptations to fundamentally different ways of life, some of which may resemble those of humans' late Miocene–early Pleistocene predecessors (i.e., those from about 11 to 2 million years ago). Laboratory and field observations, particularly of great apes, indicate that earlier researchers grossly underestimated the intelligence, cognitive abilities, and sensibilities of nonhuman primates and perhaps also those of Pliocene–early Pleistocene hominins (i.e., those from about 5.3 to 2 million years ago), who left few archaeological clues to their behaviour.

Primates

Primates are a group of mammals that includes monkeys, apes, lemurs, bushbabies, and lorises. Humans are primates, too. All primates have a large brain compared to their body size. They are also good climbers, with strong arms and legs, and long, grasping fingers and toes. The eyes of primates face forward to help them judge distance accurately.

Although there are some notable variations between some primate groups, they share several anatomic and functional characteristics reflective of their common ancestry. When compared with body weight, the primate brain is larger than that of other terrestrial mammals, and it has a fissure unique to primates (the Calcarine sulcus) that separates the first and second visual areas on each side of the brain. Whereas all other mammals have claws or hooves on their digits, only primates have flat nails. Some primates do have claws, but even among these there is a flat nail on the big toe (hallux). In all primates except humans, the hallux diverges from the other toes and together with them forms a pincer capable of grasping objects such as branches. Not all primates have similarly dextrous hands; only the catarrhines (Old World monkeys, apes, and humans) and a few of the lemurs and lorises have an opposable thumb. Primates are not alone in having grasping feet, but as these occur in many other arboreal mammals (e.g., squirrels and opossums), and as most present-day primates are arboreal, this characteristic suggests that they evolved from an ancestor that was arboreal. So too does primates' possession of specialized nerve endings (Meissner's corpuscles) in the hands and feet that increase tactile sensitivity. As far as is known, no other placental mammal has them. Primates possess dermatoglyphics (the skin ridges responsible for fingerprints), but so do many other arboreal mammals.

Old World and New World monkeys.

The eyes face forward in all primates so that the eyes' visual fields overlap. Again, this feature is not by any means restricted to primates, but it is a general feature seen among predators. It has been proposed, therefore, that the ancestor of the primates was a predator, perhaps insectivorous. The optic fibres in almost all mammals cross over (decussate) so that signals from one eye are interpreted on the opposite side of the brain, but, in some primate species, up to 40 percent of the nerve fibres do not cross over.

Primate teeth are distinguishable from those of other mammals by the low, rounded form of the molar and premolar cusps, which contrast with the high, pointed cusps or elaborate ridges of other placental mammals. This distinction makes fossilized primate teeth easy to recognize.

Fossils of the earliest primates date to the Early Eocene Epoch (56 million to 40 million years ago) or perhaps to the Late Paleocene Epoch (59 million to 56 million years ago). Though they began as an arboreal group, and many (especially the platyrrhines, or New World monkeys) have remained thoroughly arboreal, many have become at least partly terrestrial, and many have achieved high levels of intelligence.

Species of Lemurs (suborder Strepsirrhini).

Cetology

Cetology is the branch of marine mammal science the branch of marine mammal science that studies the approximately eighty species of whales, dolphins, and porpoise in the scientific order Cetacea. Cetologists, or those who practice cetology, seek to understand and explain cetacean evolution, distribution, morphology, behavior, community dynamics, and other topics.

A researcher fires a biopsy dart at an orca. The dart will remove a small piece of the whale's skin and bounce harmlessly off the animal.

Studying Cetaceans

Humpback whales often have distinct markings that enable scientists to identify individuals. Studying cetaceans presents numerous challenges. Cetaceans only spend 10% of their time on the surface, and all they do at the surface is breathe. There is very little behavior seen at the surface. It is also impossible to find any signs that an animal has been in an area. Cetaceans do not leave tracks that can be followed. However, the dung of whales often floats and can be collected to tell important information about their diet and about the role they have in the environment. Many times cetology consists of waiting and paying close attention.

Cetologists use equipment including hydrophones to listen to calls of communicating animals, binoculars and other optical devices for scanning the horizon, cameras, notes, and a few other devices and tools.

An alternative method of studying cetaceans is through examination of dead carcasses that wash up on the shore. If properly collected and stored, these carcasses can provide important information that is difficult to obtain in field studies.

Identifying Individuals

In recent decades, methods of identifying individual cetaceans have enabled accurate population counts and insights into the life cycles and social structures of various species.

One such successful system is photo-identification. This system was popularized by Michael Bigg, a pioneer in modern orca (killer whale) research. During the mid-1970s, Bigg and Graeme Ellis photographed local orcas in the British Columbian seas. After examining the photos, they realized they could recognize certain individual whales by looking at the shape and condition of the dorsal fin, and also the shape of the saddle patch. These are as unique as a human fingerprint; no one animal's looks is exactly like another's. After they could recognize certain individuals, they found that the animals travel in stable groups called pods.

Cetacean

Cetaceans as Mammals

Cetaceans are mammals. Mammals are the class (Mammalia) of vertebrate animals characterized by the presence of hair and mammary glands, which in females produce milk for the nourishment of young. As mammals, cetaceans have characteristics that are common to all mammals: They are warm-blooded, breathe in air utilizing lungs, bear their young alive and suckle them on their own milk, and have hair.

Whales, like mammals, also have a diaphragm, a muscle below the rib cage that aids breathing and it is a prehepatic diaphragm, meaning it is front of the liver. Mammals also are the only vertebrates with a single bone in the lower jaw.

Another way of discerning a cetacean from a fish is by the shape of the tail. The tail of a fish is vertical and moves from side to side when the fish swims. The tail of a cetacean has two divisions, called flukes, which are horizontally flattened and move up and down, as cetaceans' spines bend in the same manner as a human spine.

Whales have very limited hair in isolated areas, thus reducing drag in the water. Instead, they maintain internal temperatures with a thick layer of blubber (vascularized fat).

The flippers of cetaceans, as modified front limbs, show a full complement of arm and hand bones, albeit compressed in length.

The range in body size is greater for the cetaceans than for any other mammalian order.

Types of Cetaceans

Cetaceans are divided into two major suborders: Mysticeti (baleen whales) and Odontoceti (toothed whales, including whales, dolphins, and porpoises).

Mysticeti. The baleen whales (Mysticeti) are characterized by the baleen, a sieve-like structure in the upper jaw made of the tough, structural protein keratin. The baleen is used to filter plankton from the water. The mysticete skull has a bony, large, broad, and flat upper jaw, that is placed back under the eye region. They are characterized by two blowholes. Baleen whales are the largest whales. The families of baleen whales include the Balaenopteridae (humpback whales, fin whales, Sei Whale, and others), the Balaenidae (right and bowhead whales), the Eschrichtiidae (gray whale), and the Neobalaenidae (pygmy right whales), among others. The Balaenopteridae family (rorquals) also includes the Blue Whale, the world's largest animal.

Odontoceti. The toothed whales (Odontoceti) have teeth and prey on fish, squid, or both. This suborder includes dolphins and porpoises as well as whales. In contrast with the mysticete skull, the main bones of the odontocete upper jaw thrust upward and back over the eye sockets. Toothed whales have only one blowhole. An outstanding ability of this group is to sense their surrounding environment through echolocation. In addition to numerous species of dolphins and porpoises, this suborder includes the Beluga whale and the sperm whale, which may be the largest toothed animal to ever inhabit Earth. Families of toothed whales include, among others, the Monodontidae (belugas, narwhals), Kogiidae (Pygmy and dwarf sperm whales), Physteridae (sperm whale), and Ziphidae (beaked whales).

The terms whale, dolphin, and porpoise are used inconsistently and often create confusion. Members of Mysticeti are all considered whales. However, distinguishing whales, dolphins, and porpoises among the Odontoceti is difficult. Body size is useful, but not a definitive distinction, with those cetaceans greater than 9ft (2.8m) generally called whales; however, some "whales" are not that large and some dolphins can grow larger. Scientifically, the term porpoise should be reserved for members of the family Phocoenidae, but historically has been often applied in common venacular to any small cetacean. There is no strict definition of the term dolphin.

Aquatic Mammal

Aquatic and semiaquatic mammals are a diverse group of mammals that dwell partly or entirely in bodies of water. They include the various marine mammals who dwell in oceans, as well as various freshwater species, such as the European otter. They are not a taxon and are not unified by any distinct biological grouping, but rather their dependence on and integral relation to aquatic ecosystems. The level of dependence on aquatic life varies greatly among species. Among freshwater taxa, the Amazonian manatee and river dolphins are completely aquatic and fully dependent on aquatic ecosystems. Semiaquatic freshwater taxa include the Baikal seal, which feeds underwater but rests, molts, and breeds on land; and the capybara and hippopotamus which are able to venture in and out of water in search of food.

An Amazon river dolphin (*Inia geoffrensis*), a member
of the infraorder Cetacea of the order Cetartiodactyla.

Mammal adaptation to an aquatic lifestyle vary considerably between species. River dolphins and manatees are both fully aquatic and therefore are completely tethered to a life in the water. Seals are semiaquatic; they spend the majority of their time in the water, but need to return to land for important activities such as mating, breeding and molting. In contrast, many other aquatic mammals, such as hippopotamus, capybara, and water shrews, are much less adapted to aquatic living. Likewise, their diet ranges considerably as well, anywhere from aquatic plants and leaves to small fish and crustaceans. They play major roles in maintaining aquatic ecosystems, beavers especially.

Aquatic mammals were the target for commercial industry, leading to a sharp decline in all populations of exploited species, such as beavers. Their pelts, suited for conserving heat, were taken during the fur trade and made into coats and hats. Other aquatic mammals, such as the Indian rhinoceros, were targets for sport hunting and had a sharp population decline in the 1900s. After it was made illegal, many aquatic mammals became subject to poaching. Other than hunting, aquatic mammals can be killed as bycatch from fisheries, where they become entangled in fixed netting and drown or starve. Increased river traffic, most notably in the Yangtze river, causes collisions between fast ocean vessels and aquatic mammals, and damming of rivers may land migratory aquatic mammals in unsuitable areas or destroy habitat upstream. The industrialization of rivers led to the extinction of the Chinese river dolphin, with the last confirmed sighting in 2004.

Marine Mammal

Marine mammals are aquatic mammals that rely on the ocean and other marine ecosystems for their existence. They include animals such as seals, whales, manatees, sea otters and polar bears. They do not represent a distinct taxon or systematic grouping, and are unified only by their reliance on the marine environment for feeding.

A humpback whale (*Megaptera novaeangliae*), a member of infraorder Cetacea of the order Cetartiodactyla.

A leopard seal (*Hydrurga leptonyx*).

Marine mammal adaptation to an aquatic lifestyle varies considerably between species. Both cetaceans and sirenians are fully aquatic and therefore are obligate water dwellers. Seals and sea-lions are semiaquatic; they spend the majority of their time in the water but need to return to land for important activities such as mating, breeding and molting. In contrast, both otters and the polar bear are much less adapted to aquatic living. Their diet varies considerably as well; some may eat zooplankton, others may eat fish, squid, shellfish, sea-grass and a few may eat other mammals. While the number of marine mammals is small compared to those found on land, their roles in various ecosystems are large, especially concerning the maintenance of marine ecosystems, through processes including the regulation of prey populations. This role in maintaining ecosystems makes them of particular concern as 23% of marine mammal species are currently threatened.

Marine mammals were first hunted by aboriginal peoples for food and other resources. Many were also the target for commercial industry, leading to a sharp decline in all populations of exploited species, such as whales and seals. Commercial hunting led to the extinction of Steller's sea cow, sea mink, Japanese sea lion and the Caribbean monk seal. After commercial hunting ended, some species, such as the gray whale and northern elephant seal, have rebounded in numbers; conversely, other species, such as the North Atlantic right whale, are critically endangered. Other than hunting, marine mammals can be killed as bycatch from fisheries, where they become entangled in fixed netting and drown or starve. Increased ocean traffic causes collisions between fast ocean vessels and large marine mammals. Habitat degradation also threatens marine mammals and their ability to find and catch food. Noise pollution, for example, may adversely affect echolocating mammals, and the ongoing effects of global warming degrade Arctic environments.

Hippology

Hippology is the study of the horse.

Horse

The horse (Equus ferus caballus) is one of two extant subspecies of Equus ferus. It is an odd-toed ungulate mammal belonging to the taxonomic family Equidae. The horse has evolved over the past 45 to 55 million years from a small multi-toed creature, Eohippus, into the large, single-toed animal of today. Humans began domesticating horses around 4000 BC, and their domestication is believed to have been widespread by 3000 BC. Horses in the subspecies caballus are domesticated, although some domesticated populations live in the wild as feral horses. These feral populations are not true wild horses, as this term is used to describe horses that have never been domesticated, such as the endangered Przewalski's horse, a separate subspecies, and the only remaining true wild horse. There is an extensive, specialized vocabulary used to describe equine-related concepts, covering everything from anatomy to life stages, size, colors, markings, breeds, locomotion, and behavior.

Horses' anatomy enables them to make use of speed to escape predators and they have a well-developed sense of balance and a strong fight-or-flight response. Related to this need to flee from predators in the wild is an unusual trait: horses are able to sleep both standing up and lying down, with younger horses tending to sleep significantly more than adults. Female horses, called mares, carry their young for approximately 11 months, and a young horse, called a foal, can stand and run shortly following birth. Most domesticated horses begin training under saddle or in harness between the ages of two and four. They reach full adult development by age five, and have an average lifespan of between 25 and 30 years.

Horse breeds are loosely divided into three categories based on general temperament: spirited "hot bloods" with speed and endurance; "cold bloods", such as draft horses and some ponies, suitable for slow, heavy work; and "warmbloods", developed from crosses between hot bloods and cold bloods, often focusing on creating breeds for specific riding purposes, particularly in Europe. There are more than 300 breeds of horse in the world today, developed for many different uses.

Horses and humans interact in a wide variety of sport competitions and non-competitive recreational pursuits, as well as in working activities such as police work, agriculture, entertainment, and therapy. Horses were historically used in warfare, from which a wide variety of riding and driving techniques developed, using many different styles of equipment and methods of control. Many products are derived from horses, including meat, milk, hide, hair, bone, and pharmaceuticals extracted from the urine of pregnant mares. Humans provide domesticated horses with food, water, and shelter, as well as attention from specialists such as veterinarians and farriers.

Cynology

Cynology is the study of anything related to domestic dogs. It includes such subjects as the evolution of the dog, and its anatomy and physiology. It covers all aspects of general dog care such as feeding, training and breeding.

Dog

The domestic dog (Canis lupus familiaris when considered a subspecies of the wolf or Canis familiaris when considered a distinct species) is a member of the genus Canis (canines), which forms part of the wolf-like canids, and is the most widely abundant terrestrial carnivore. The dog and the extant gray wolf are sister taxa as modern wolves are not closely related to the wolves that were first domesticated, which implies that the direct ancestor of the dog is extinct. The dog was the first species to be domesticated and has been selectively bred over millennia for various behaviors, sensory capabilities, and physical attributes.

Their long association with humans has led dogs to be uniquely attuned to human behavior and they are able to thrive on a starch-rich diet that would be inadequate for other canid species. Dogs vary widely in shape, size and colors. They perform many roles for humans, such as hunting, herding, pulling loads, protection, assisting police and military, companionship and, more recently, aiding disabled people and therapeutic roles. This influence on human society has given them the sobriquet of "man's best friend".

Dogs show great morphological variation.

The term *dog* typically is applied both to the species (or subspecies) as a whole, and any adult male member of the same. An adult female is a *bitch*. An adult male capable of reproduction is a *stud*. An adult female capable of reproduction is a *brood bitch*, or *brood mother*. Immature males or females (that is, animals that are incapable of reproduction) are *pups* or *puppies*. A group of pups from the same gestation period is called a *litter*. The father of a litter is a *sire*. It is possible for one litter to have multiple sires. The mother of a litter is a *dam*. A group of any three or more adults is a *pack*.

Felinology

Felinology is the study of cats. Felinology is concerned with studying the anatomy, genetics, physiology, and breeding of domestic and wild cats.

Felidae

The Felidae family is a part of the order Carnivora within the mammals (Class Mammalia). Members of the family are called cats or felids, and sometimes felines, although this latter term is more precisely used for members of the subfamily Felinae. They number about 41 species including large animals such as the lion (Panthera leo) and the tiger (Panthera tigris), as well as smaller ones such as the bobcat (Lynx rufus) and the domestic cat (Felis catus).

Some cats present a threat to domestic animals, even to humans in the case of the big cats. For such reasons, as well as for sport and fur, cats have been a target of hunters and trappers. Because of such hunting, combined with loss of suitable habitat and other causes, most species are now considered to be be endangered in the wild.

Yet, as with all animals, species within Felidae provide a larger benefit to the ecosystem and to humans while pursuing their own individual purpose of survival, maintenance, and reproduction. Ecologically, as apex predators at the top of food chains, they play an important role in keeping populations of prey species under control, and thus aiding the balance of nature. Cats also are a large part of people's fascination with nature, with their appearance and behaviors creating aesthetic delight and wonder, whether being viewed in nature, in zoos, or in media. The domestic cat is a popular companion in people's homes and has historically benefited their hosts by killing rodents. Cats are admired for their beauty, their grace, and their seemingly mysterious ways, and they have often been featured in art and legend.

Characteristics

Like most other members of Carnivora (carnivores), cats mainly get food by killing and eating other animals. They are more strictly carnivorous (meat eating) than most other carnivore families. Bobcats and some other cat species supplement their meat diet with fruit. The teeth of cats are well suited to their diet, with long canines for gripping prey and blade-like molars for cutting flesh (Voelker 1986).

All cats walk on four feet, in a digitigrade manner—that is on their toes. Their hind legs are longer and stronger than their fore legs, which gives them strength for sprinting and leaping, but not stamina for long distance running. Cats have claws on their feet that are used for gripping prey, for fighting, and for climbing. In all cats except the cheetah *(Acinonyx jubatus)*, the fishing cat *(Prionailurus viverrinus)*, and the flat-headed cat *(Prionailurus planiceps)*, the claws can be retracted into the toes, which helps keep them sharp.

Most cats have a long fur-covered tail, which is used for balance in running and leaping, and sometimes for warmth. The bodies of all cats, except some breeds of domestic cat, are covered with thick fur. In most species, this is spotted or striped.

Cats have very keen senses, especially their vision. Their eyes are large and are well suited for seeing in low levels of light. Most cats hunt at night or in the late evening and early morning. They can not, however, see in total darkness.

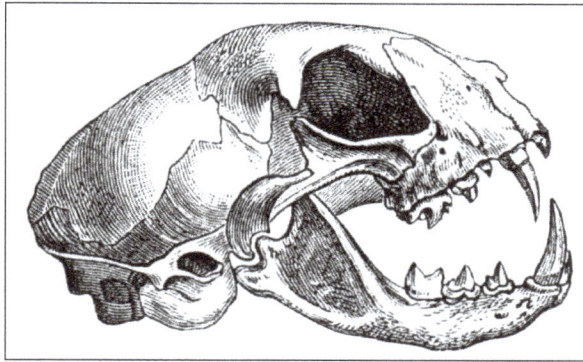
Wildcat skull.

With a few exceptions, most notably lions, cats live most of their lives alone. Male and female cats come together to mate, which in most species happens once a year. The young are born in litters of one to six. They are cared for by their mother for several months, until they are mature and experienced enough to live on their own.

The cat family is usually divided into the "big cats" of the subfamily Pantherinae and the "small cats" of the subfamily Felinae. The largest cat is the tiger (the subspecies Siberian Tiger), which can weigh as much as 250 kg (550 lb). The smallest cat is the black-footed cat of southern Africa *(Felis nigripes)*, which weighs about 1.5 - 2.75 kg (3.3 - 6 lb). Some "small cats," for instance the cougar *(Puma concolor)*, can weigh as much or more as some of the "big cats." Ligers, crossbreeds between male lions *(Panthera leo)* and female tigers *(Panthera tigris)*, which can weigh more than 450 kg (1000 lbs), are the largest cats in the world if hybrids are included. (A similar, but smaller hybrid, the offspring of a male tiger and a female lion, is called a *tigon.*

Role in Nature

Fishing cat.

Cats are found in the wild in most land environments, on all continents except Antarctica and Australia (except for feral domestic cats). Some species of cats are native to tropical rainforests, grasslands, deserts, temperate forests, and high mountains.

Cats, like other predators, play an important role in keeping the populations of prey species under control so that excessive damage to plants is avoided and the balance of nature is preserved.

Among the animals preyed upon by cats are rodents, birds, reptiles, and in the case of the larger species hoofed animals. Lions have been known to prey on African elephants, the world's largest land animal. The fishing cat of Southeast Asia is a good swimmer and eats mainly fish.

Feral cats—domestic cats that have returned to the wild—are a problem in many areas of the world and have caused the extinction of some native species of birds and mammals.

Cats and Humans

To humans, the most important cat is the domestic cat, which is descended from the wild cat *(Felis sylvestris)*. Since the time of ancient Egypt, domestic cats, or their wild cat ancestors, have shared the homes of humans and have greatly benefited their hosts by killing destructive rodents. Domestic cats are now one of the most popular pets and are common all over the world.

Although most cat species are beneficial to humans because of their role in nature, some present a threat to domestic animals and, in the case of some of the big cats, to humans themselves. For this reason, as well as for their fur and for sport, cats have been hunted and trapped. Most cat species are now considered to be endangered in the wild.

In general, cats are admired by humans for their beauty, their grace, and their seemingly mysterious ways. They have often been featured in art and in legends and folktales. Cats are often used in advertising and as mascots for sports teams and military units.

Fossil Felids

The oldest known true felid (Proailurus) lived in the Oligocene and Miocene eras. During the Miocene, it gave way to Pseudaelurus. Pseudaelurus is believed to be the latest common ancestor of the two extant subfamilies, Pantherinae and Felinae, and the extinct subfamily, Machairodontinae. This group, better known as the sabertooth cats, became extinct in the Late Pleistocene era. It includes the genera Smilodon, Machairodus, Dinofelis, and Homotherium.

References

- Primatology, science: britannica.com, Retrieved 30 March, 2019

- Powell, J.; Kouadio, A. (2008). "Trichechus senegalensis". IUCN Red List of Threatened Species. Version 2008. International Union for Conservation of Nature. Retrieved 26 June 2016

- Primates, animals-and-nature: dkfindout.com, Retrieved 1 April, 2019

- Meng Chen, Gregory Philip Wilson, A multivariate approach to infer locomotor modes in Mesozoic mammals, Article in Paleobiology 41(02) · February 2015 DOI: 10.1017/pab.2014.14

- General-considerations, primate-mammal, animal: britannica.com, Retrieved 2 May, 2019

- Kingdon, Jonathan; Happold, David; Butynski, Thomas; Hoffmann, Michael; Happold, Meredith; Kalina, Jan (2013). Mammals of Africa. 1. New York: Bloomsbury Publishing. P. 211. ISBN 978-1-4081-2251-8. OCLC 822025146

- Felidae, entry: newworldencyclopedia.org, Retrieved 3 June, 2019

- Edwards, Holly H.; Schnell, Gary D. (2001). "Body Length, Swimming Speed, Dive Duration, and Coloration of the Dolphin Sotalia fluviatilis (Tucuxi) in Nicaragua" (PDF). Caribbean Journal of Science. 37: 271–298

Anatomy and Physiology of Mammals

3

- **Mammal Anatomy and Physiology**
- **Primate Anatomy and Physiology**
- **Cetacea Anatomy and Physiology**
- **Equine Anatomy and Physiology**
- **Dog Anatomy and Physiology**
- **Felidae Anatomy and Physiology**

Anatomy and physiology are the branches of science that deal with bodily structure and functioning of living organisms respectively. It includes the study of mammals such as cetacea, equine, felidae, etc. This chapter delves into these aspects associated with mammalogy for providing an in-depth understanding of the subject.

Mammal Anatomy and Physiology

Mammals all look very different. Inside, however, they have the same basic bony skeleton. Mammals have a backbone, just like birds, fish, amphibians, and reptiles. They belong to a group called vertebrates, which are animals with a backbone. The skeleton provides a framework for a mammal's body and protects the inner organs. Attached to the skeleton are muscles, which pull on the bones and allow the animals to move. The skeletons of different kinds of mammals allow them to perform a wide range of movements—including running, swimming, climbing, and flying.

Skin and Hair

The skin of mammals is constructed of two layers, a superficial nonvascular epidermis and an inner layer, the dermis, or corium. The two layers interlock via fingerlike projections (dermal papillae), consisting of sensitive vascular dermis projecting into the epidermis. The outermost layers of the epidermis are cornified (impregnated with various tough proteins), and their cells are enucleate (lacking cell nuclei). The epidermis is composed of flattened cells in layers and is the interface between the individual and the environment. Its primary function is defensive, and it is cornified to resist abrasion. The surface of the skin is coated with lipids and organic salts, the so-called "acid mantle," which is thought to possess antifungal and antibacterial properties. Deep in the epidermis is an electronegative (electron-attracting) layer, a further deterrent to foreign organic or ionic agents.

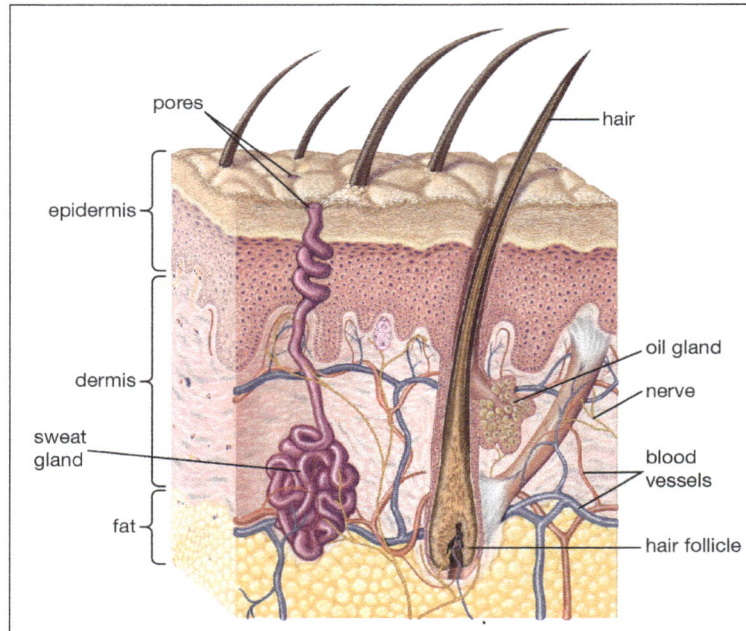

Skin a cross section of mammalian skin and its underlying structures.

The dermis lies beneath the epidermis and nourishes it. The circulation of the dermis is variously developed in mammals, but it is typically extensive, out of proportion to the nutritional needs of the tissue. Its major role is to moderate body temperature and blood pressure by forming a peripheral shunt, an alternate route for the blood. Also in the dermis are sensory nerve endings to alert the individual to pressure (touch), heat, cold, and pain. In general, skin bearing hairs has few or no specialized sensory endings. The sensation of touch on hairy skin in humans depends on stimulation of the nerve fibres associated with the hairs. However, hairless skin, such as the lips and fingertips, has specialized endings.

Hair is derived from an invagination (pocketing) of the epidermis termed a follicle. Collectively, the hair is called the pelage. The individual hair is a rod of keratinized cells that may be cylindrical or more or less flattened. Keratin is a protein also found in claws and nails. The inner medulla of the hair is hollow and contains air; in the outer cortex layer there are frequently pigment granules. Associated with the hair follicle are nerve endings and a muscle, the arrector pili. The latter allows the erection of individual hairs to alter the insulative qualities of the pelage. The follicle also gives rise to sebaceous glands that produce sebum, a substance that lubricates the hair.

Most mammals have three distinct kinds of hairs. Guard hairs protect the rest of the pelage from abrasion and frequently from moisture, and they usually lend a characteristic colour pattern. The thicker underfur is primarily insulative and may differ in colour from the guard hairs. The third common hair type is the vibrissa, or whisker, a stiff, typically elongate hair that functions in tactile sensation. Hairs may be further modified to form rigid quills. The "horn" of the rhinoceros is composed of a fibrous keratin material derived from hair. Examples of keratinized derivatives of the integument other than hair are horns, hooves, nails, claws, and baleen.

Even though the primary function of the skin is defensive, it has been modified in mammals to serve such diverse functions as thermoregulation and nourishment of young. Secretions of sweat glands promote cooling due to evaporation at the surface of the body, and mammary glands are a type of apocrine gland (that is, a sweat gland associated with hair follicles).

In certain groups (primates in particular) the skin of the face is under intricate muscular control, and movements of the skin express and communicate emotion. In many mammals the colour and pattern of the pelage are important in communicative behaviour. Patterns may be startling (dymantic), as seen in the mane of the male lion or hamadryas baboon, warning (sematic), as seen in the bold pattern of skunks, or concealing (cryptic), perhaps the most common adaptation of pelage colour.

Hair has been secondarily lost or considerably reduced in some kinds of mammals. In adult cetaceans insulation is provided by thick subcutaneous fat deposits, or blubber, with hair limited to a few stiff vibrissae about the mouth. The bare skin is one of a number of features that contribute to the remarkably advanced hydrodynamics of locomotion in the group. Some burrowing (fossorial) mammals also tend toward reduction of the hair. This is shown most strikingly by the sand rats of northeastern Africa, but considerable loss of hair has also occurred in some species of pocket gophers. Hair may also be lost on restricted areas of the skin, as from the face in many monkeys or the buttocks of mandrills, and may be sparse on elephants and such highly modified species as pangolins and armadillos.

Continuous growth of hair (indeterminate), as seen on the heads of humans, is rare among mammals. Hairs with determinate growth are subject to wear and must be replaced period-ically—a process termed molt. The first coat of a young mammal is referred to as the juvenal pelage, which typically is of fine texture like the underfur of adults and is replaced by a post-juvenile molt. Juvenal pelage is succeeded either directly by adult pelage or by the subadult pelage, which in some species is not markedly distinct from that of the adult. Once this pel-age has been acquired, molting continues to recur at intervals, often annually or semiannually and sometimes more frequently. The pattern of molt typically is orderly, but it varies widely between species. Some mammals apparently molt continuously, with a few hairs at a time re-placed throughout the year.

Teeth

Specialization in food habits has led to profound dental changes. The primitive mammalian tooth had high, sharp cusps and served to tear flesh or crush chitinous material (primarily the exoskel-etons of terrestrial arthropods, such as insects). Herbivores tend to have specialized cheek teeth with complex patterns of contact (occlusion) and various ways of expanding the crowns of the teeth and circumventing the problem of wear. Omnivorous mammals, such as bears, pigs, and humans, tend to have molars with low, rounded cusps, termed bunodont.

A prime example of convergence in conjunction with dietary specialization is seen in those mam-mals adapted to feeding on ants and termites, a specialization generally termed myrmecophagy ("ant eating"). Trends frequently associated with myrmecophagy include strong claws, an elongate rounded skull, a wormlike extensible tongue, marked reduction in the mandible (lower jaw), and loss or extreme simplification of the teeth (dentition). This habit has led to remarkably similar morphology among animals as diverse as the echidna (a monotreme), the numbat (a marsupial), the anteater (a xenarthran), the aardvark (a tubulidentate), and the pangolin (a pholidotan).

A yawning African lion (Panthera leo) showing its long canine teeth.

Specialized herbivores evolved early in mammalian history. The extinct multituberculates were the earliest mammalian herbivores and have the longest evolutionary history, lasting more than 100 million years from 178 million to 50 million years ago. Multituberculate fossils, such as those of Ptilodus, dated to the Paleocene Epoch (66–56 million years ago) of North America, have been found on all continents. Similarities in teeth not due to common ancestry have occurred widely in herbivorous groups. Most herbivores have incisors modified for nipping or gnawing, have lost teeth with the resultant development of a gap (diastema) in the tooth row, and exhibit some molarization (expansion and flattening) of premolars to expand the grinding surface of the cheek teeth. Rootless incisors or cheek teeth have evolved frequently, their open pulp cavity allowing continual growth throughout life. Herbivorous specializations have evolved independently in multituberculates, rodents, lagomorphs, primates, and the wide diversity of ungulate and subungulate orders.

Skeleton

The mammalian skeletal system shows a number of advances over that of lower vertebrates. The mode of ossification (bone formation) of the long bones is characteristic. In lower vertebrates each long bone has a single centre of ossification (the diaphysis), and replacement of cartilage by bone proceeds from the centre toward the ends, which may remain cartilaginous, even in adults. In mammals secondary centres of ossification (the epiphyses) develop at the ends of the bones. Growth of bones occurs in zones of cartilage between diaphysis and epiphyses. Mammalian skeletal growth is termed determinate, for once the actively growing zone of cartilage has been obliterated, growth in length ceases. As in all bony vertebrates, of course, there is continual renewal of bone throughout life. The advantage of epiphyseal ossification lies in the fact that the bones have strong articular (joint-related) surfaces before the skeleton is mature. In general, the skeleton of the adult mammal has less structural cartilage than does that of a reptile.

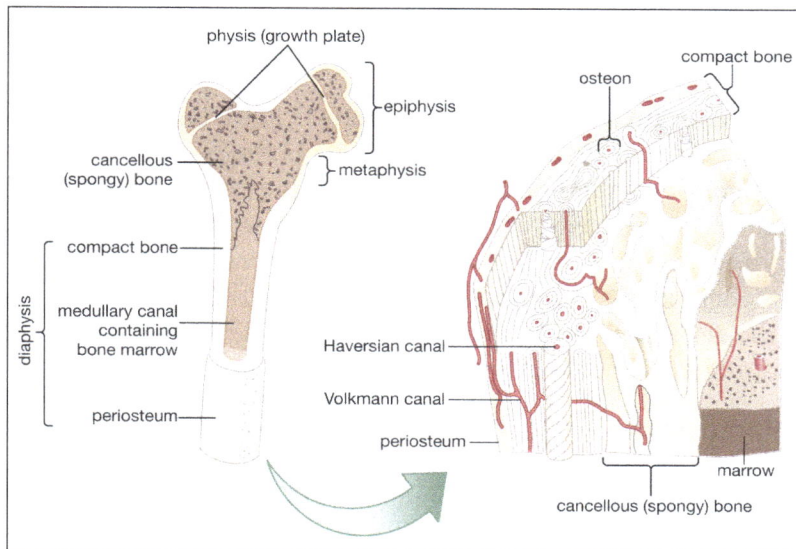

Internal structure of a human long bone, with a magnified cross section of the interior. The central tubular region of the bone, called the diaphysis, flares outward near the end to form the metaphysis, which contains a largely cancellous, or spongy, interior. At the end of the bone is the epiphysis, which in young people is separated from the metaphysis by the physis, or growth plate. The periosteum is a connective sheath covering the outer surface of the bone. The Haversian system, consisting of inorganic substances arranged in concentric rings around the Haversian canals, provides compact bone with structural support and allows for metabolism of bone cells. Osteocytes (mature bone cells) are found in tiny cavities between the concentric rings. The canals contain capillaries that bring in oxygen and nutrients and remove wastes. Transverse branches are known as Volkmann canals.

The skeletal system of mammals and other vertebrates is broadly divisible functionally into axial and appendicular portions. The axial skeleton consists of the braincase (cranium) and the backbone and ribs, and it serves primarily to protect the central nervous system. The limbs and their girdles constitute the appendicular skeleton. In addition, there are skeletal elements derived from the gill arches of primitive vertebrates, collectively termed the visceral skeleton. Visceral elements in the mammalian skeleton include the jaws, the hyoid apparatus supporting the tongue, and the auditory ossicles of the middle ear. The postcranial axial skeleton in mammals generally has remained rather conservative during the course of evolution. The vast majority of mammals have seven cervical (neck) vertebrae; exceptions are sloths, with six or nine cervicals, and the sirenians with six. The anterior two cervical vertebrae are differentiated as atlas and axis. Specialized articulations of these two bones allow complex movements of the head on the trunk. Thoracic vertebrae bear ribs and are variable in number. The anterior ribs converge toward the ventral midline to articulate with the sternum, or breastbone, forming a semirigid thoracic "basket" for the protection of heart and lungs. Posterior to the thoracic region are the lumbar vertebrae, ranging from 2 to 21 in number (most frequently 4 to 7). Mammals have no lumbar ribs. There are usually 3 to 5 sacral vertebrae, but some xenarthrans have as many as 13. Sacral vertebrae fuse to form the sacrum, to which the pelvic girdle is attached. Caudal (tail) vertebrae range in number from 5 (fused elements of the human coccyx or tailbone) to 50.

The basic structure of the vertebral column is comparable throughout the Mammalia, although in many instances modifications have occurred in specialized locomotor modes to gain particular mechanical advantages. The vertebral column and associated muscles of many mammals are structurally analogous to a cantilever girder.

The skull is composite in origin and complex in function. Functionally the bones of the head are separable into the braincase and the jaws. In general, it is the head of the animal that meets the environment. The skull protects the brain and sense capsules (the parts of the skeleton that facilitate the senses of sight, hearing, taste, and smell), houses the teeth and tongue, and contains the entrance to the pharynx. Thus, the head functions in sensory reception, food acquisition, defense, respiration, and (in higher groups) communication. To serve these functions, bony elements have been recruited from the visceral skeleton, the endochondral skeleton (the parts of the skeleton that form from cartilage), and the dermal skeleton of lower vertebrates.

The skull of mammals differs markedly from that of reptiles because of the great expansion of the brain. The sphenoid bones that form the reptilian braincase form only the floor of the braincase in mammals. The side is formed in part by the alisphenoid bone, derived from the epipterygoid, a part of the reptilian palate. Dermal elements, the frontals and parietals, have come to lie deep to (beneath) the muscles of the jaw to form the dorsum of the braincase. Reptilian dermal roofing bones, lying superficial to the muscles of the jaw, are represented in mammals only by the jugal bone of the zygomatic arch, which lies under the eye.

In mammals a secondary palate is formed by processes of the maxillary bones and the palatines, with the pterygoid bones reduced in importance. The secondary palate separates the nasal passages from the oral cavity and allows continuous breathing while chewing or suckling.

Other specializations of the mammalian skull include paired articulating surfaces at the neck (occipital condyles) and an expanded nasal chamber with complexly folded turbinal bones, providing a large area for detection of odours. Eutherians have evolved bony protection for the middle ear, the auditory bulla. The development of this structure varies, although a ring-shaped (annular) tympanic bone is always present.

The bones of the mammalian middle ear are a diagnostic feature of the class. The three auditory ossicles form a series of levers that serve mechanically to increase the amplitude of sound waves reaching the tympanic membrane, or eardrum, produced as disturbances of the air. The innermost bone is the stapes, or "stirrup bone." It rests against the oval window of the inner ear. The stapes is homologous with the entire stapedial structure of reptiles, which in turn was derived from the hyomandibular arch of primitive vertebrates. The incus, or "anvil," articulates with the stapes. The incus was derived from the quadrate bone, which is involved in the jaw articulation in reptiles. The malleus, or "hammer," rests against the tympanic membrane and articulates with the incus. The malleus is the homologue of the reptilian articular bone. The mechanical efficiency of the middle ear has thus been increased by the incorporation of two bones of the reptilian jaw assemblage. In mammals the lower jaw is a single bone, the dentary, which articulates with the squamosal of the skull.

The limbs and girdles have been greatly modified with locomotor adaptations. The ancestral mammal had well-developed limbs and was five-toed. In each limb there were two distal (outer) elements (radius and ulna in the forelimb; tibia and fibula in the hind limb) and a single proximal (inner or upper) element (humerus; femur). There were nine bones in the wrist, the carpals, and seven bones in the ankle, the tarsals. The phalangeal formula (the number of phalangeal bones in each digit, numbered from inside outward) is 2-3-3-3-3 in primitive mammals; in primitive reptiles it is 2-3-4-5-3. Modifications in mammalian limbs have involved reduction, loss, or fusion of bones. Loss of the clavicle from the shoulder girdle, reduction in the number of toes, and modifications of tarsal and carpal bones are typical correlates of cursorial locomotion. Scansorial and arboreal groups tend to maintain or emphasize the primitive divergence of the thumb and hallux (the inner toe on the hind foot).

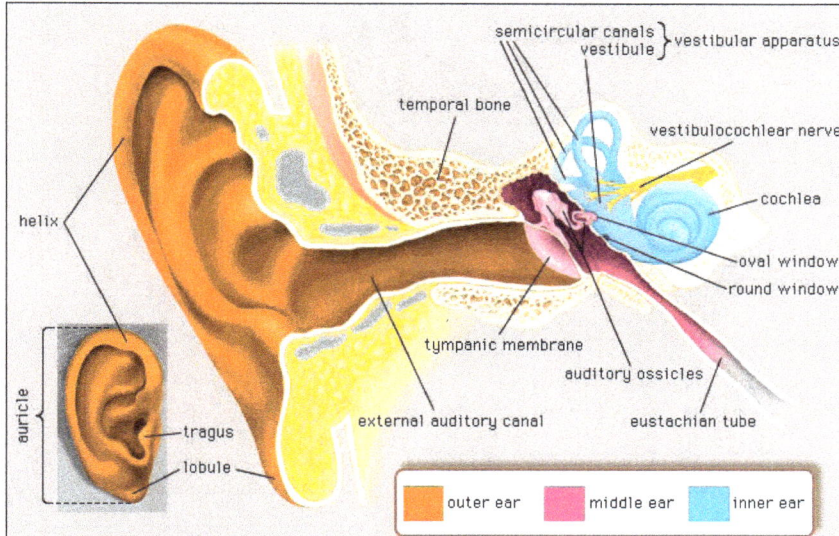

The structures of the outer, middle, and inner ear.

Centres of ossification sometimes develop in nonbony connective tissue. Such bones are termed heterotopic or sesamoid elements. The kneecap (patella) is such a bone. Another important bone of this sort, found in many kinds of mammals, is the baculum, or os penis, which occurs as a stiffening rod in the penis of such groups as carnivores, many bats, rodents, some insectivores, and many primates. The os clitoridis is a homologous structure found in females.

Muscles

The muscular system of mammals is generally comparable to that of reptiles. With changes in locomotion, the proportions and specific functions of muscular elements have been altered, but the relationships of these muscles remain essentially the same. Exceptions to this generalization are the muscles of the skin and of the jaw.

The panniculus carnosus is a sheath of dermal (skin) muscle, developed in many mammals, that allows the movement of the skin independent of the movement of deeper muscle masses. These movements function in such mundane activities as the twitching of the skin to foil insect pests and in some species also are important in shivering, a characteristic heat-producing response to thermal stress. The dermal musculature of the facial region is particularly well developed in primates and carnivores but occurs in other groups as well. Facial mobility allows expression that may be of importance in the behavioral maintenance of interspecific social structure.

The temporalis muscle is the major adductor (closer) of the reptilian jaw. In mammals the temporalis is divided into a deep temporalis proper and a more superficial masseter muscle. The temporalis attaches to the coronoid process of the mandible (lower jaw) and the temporal bone of the skull. The masseter passes from the angular process of the mandible to the zygomatic arch. The masseter allows an anteroposterior (forward-backward) movement of the jaw and is highly developed in mammals, such as rodents, for which grinding is the important function of the dentition.

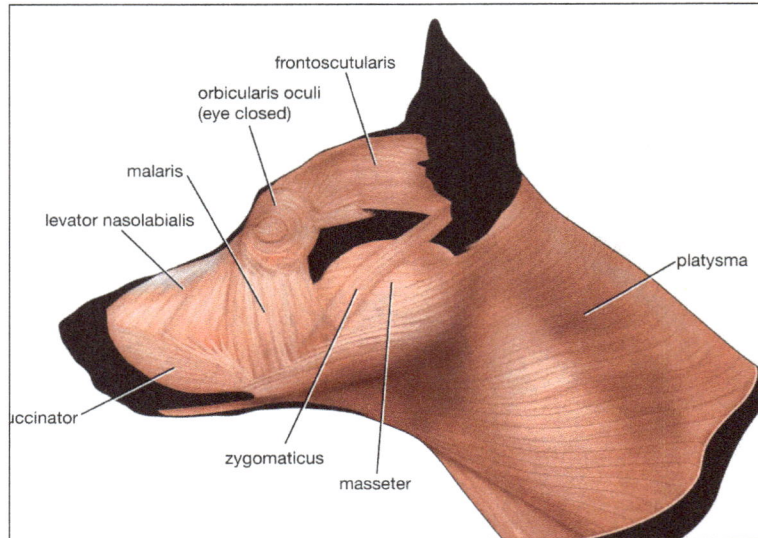

Facial musculature: Lateral view of facial musculature of a dog.

Digestive System

The alimentary canal is highly specialized in many kinds of mammals. In general, specializations of the gut accompany herbivorous habits. The intestines of herbivores are typically elongate, and the stomach may also be specialized. Subdivision of the gut allows areas of differing physiological environments for the activities of different sorts of enzymes and symbiotic bacteria, which aid the animal by breaking down certain compounds that are otherwise undigestible. In ruminant artiodactyls, such as antelopes, deer, and cattle, the stomach has up to four chambers, each with a particular function in the processing of vegetable material. A cecum is common in many herbivores. The cecum is a blind sac at the far end of the small intestine where complex compounds such as cellulose are acted upon by symbiotic bacteria. The vermiform appendix is a diverticulum of the cecum. The appendix is rich in lymphoid tissue and in many mammals is concerned with defense against toxic bacterial products.

Hares and rabbits, the sewellel, or "mountain beaver" (Aplodontia rufa), and some insectivores exhibit a phenomenon of reingestion called coprophagy, in which at intervals specialized fecal pellets are produced. These pellets are eaten and passed through the alimentary canal a second time. Where known to be present, this pattern seems to be obligatory. Reingestion primarily occurs in members of the shrew, rodent, and rabbit groups; however, the behaviour has been observed to a lesser degree in other groups, including canines and pikas. The process appears to allow the animal to absorb in the upper gut vitamins produced by the microflora of the lower gut but not absorbable there.

Excretory System

The mammalian kidney is constructed of a large number of functional units called nephrons. Each nephron consists of a distal tubule, a medial section termed the loop of Henle, a proximal tubule, and a renal corpuscle. The renal corpuscle is a knot of capillaries (glomerulus) surrounded by a sheath (Bowman's capsule). The renal corpuscle is a pressure filter, relying on blood pressure to remove water, ions, and small organic molecules from the blood. Some of the material removed is waste, but some is of value to the organism. The filtrate is sorted by the tubules, and water and needed solutes are resorbed. Resorption is both passive (osmotic) and active (based on ion transport systems). The distal convoluted tubules drain into collecting tubules, which in turn empty into the calyces, or branches, of the renal pelvis, the expanded end of the ureter. The pressure-pump nephron of mammals is so efficient that the renal portal system of lower vertebrates has been completely lost. Mammalian kidneys show considerable variety in structure, relative to the environmental demands on a given species. In particular, desert rodents have long loops of Henle and are able to resorb much water and to excrete a highly concentrated urine. Urea is the end product of protein metabolism in mammals, and excretion is therefore called ureotelic.

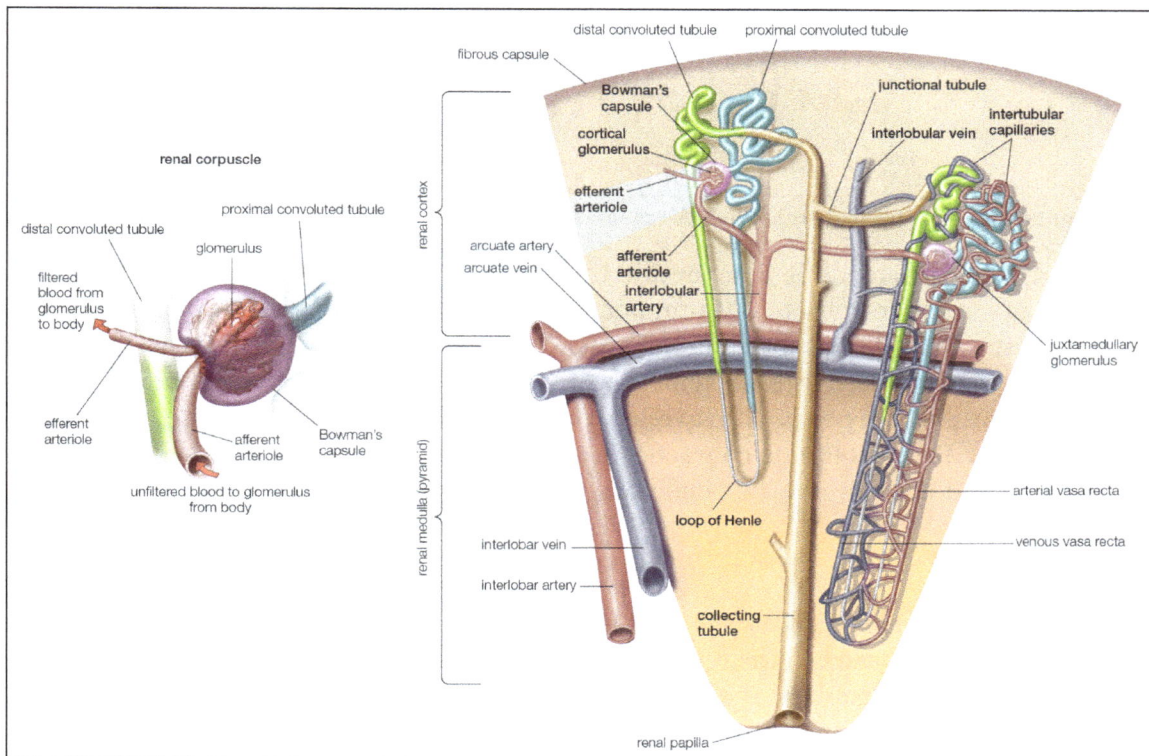

Each nephron of the kidney contains blood vessels and a special tubule. As the filtrate flows through the tubule of the nephron, it becomes increasingly concentrated into urine. Waste products are transferred from the blood into the filtrate, while nutrients are absorbed from the filtrate into the blood.

Reproductive System

The Male Tract

The testes of mammals descend from the abdominal cavity to lie in a compartmented pouch termed the scrotum. In some species the testes are permanently scrotal, and the scrotum is sealed off from the general abdominal cavity. In other species the testes migrate to the scrotum only during the breeding season. It is thought that the temperature of the abdominal cavity is too high to allow spermatogenesis; the scrotum allows cooling of the testes.

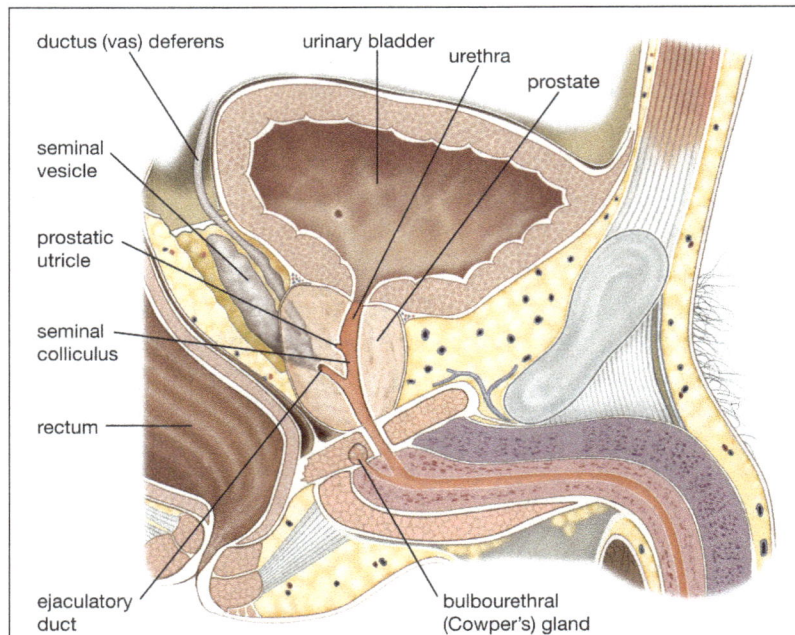

Sagittal section of the male reproductive organs, showing the prostate gland and seminal vesicles.

The transport of spermatozoa is comparable to that in reptiles, relying on ducts derived from urinary ducts of earlier vertebrates. Mammalian specialities are the bulbourethral (or Cowper's) glands, the prostate gland, and the seminal vesicle or vesicular gland. Each of these glands adds secretions to the spermatozoa to form semen, which passes from the body via a canal (urethra) in the highly vascular, erectile penis. The tip of the penis, the glans, may have a complex morphology and has been used as a taxonomic character in some groups. The penis may be retracted into a sheath along the abdomen or may be pendulous, as in bats and many primates.

The Female Tract

The structure of the female reproductive tract is variable. Four types of uterus are generally recognized among placentals, based on the relationship of the uterine horns (branches). A duplex uterus characterizes rodents and rabbits; the uterine horns are completely separated and have separate cervices opening into the vagina. Carnivores have a bipartite uterus, in which the horns are largely separate but enter the vagina by a single cervix. In the bicornate uterus, typical of many ungulates, the horns are distinct for less than half their length; the lower part of the uterus is a common chamber, the body. Higher primates have a simplex uterus in which there is no separation between the horns and thus a single chamber.

The uterus is an inverted pear-shaped muscular organ of the female reproductive system, located between the bladder and the rectum. It functions to nourish and house the fertilized egg until the unborn child is ready to be delivered.

The female reproductive tract of marsupials is termed didelphous; the vagina is paired, as are oviducts and uteri. In primitive marsupials there are paired vaginae lateral to the ureters. In more advanced groups, such as kangaroos, the lateral vaginae persist and conduct the migration of spermatozoa, but a medial "pseudovagina" functions as the birth canal.

Monotremes have paired uteri and oviducts, which empty into a urogenital sinus (cavity) as fluid wastes do. The sinus passes into the cloaca, a common receptacle for reproductive and excretory products.

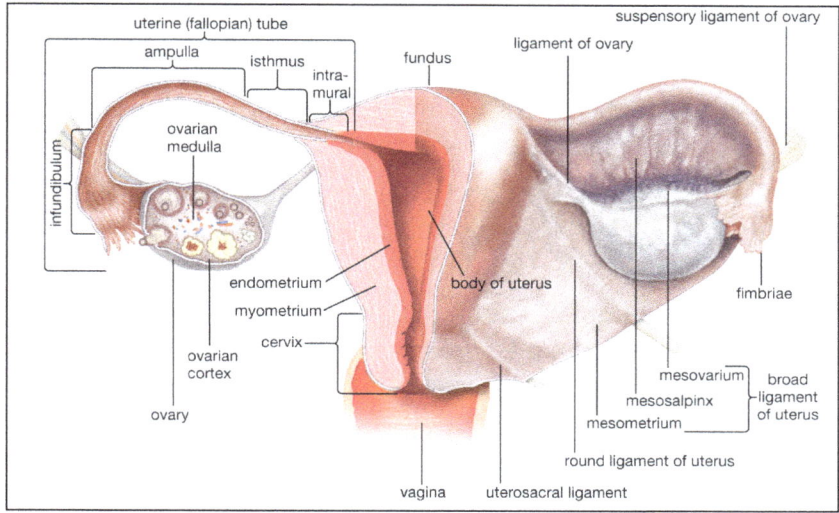

Uterus.

Circulatory System

In mammals, as in birds, the right and left ventricles of the heart are completely separated, so that pulmonary (lung) and systemic (body) circulations are completely independent. Oxygenated blood arrives in the left atrium from the lungs and passes to the left ventricle, whence it is forced through the aorta to the systemic circulation. Deoxygenated blood from the tissues returns to the right atrium via a large vein, the vena cava, and is pumped to the pulmonary capillary bed through the pulmonary artery.

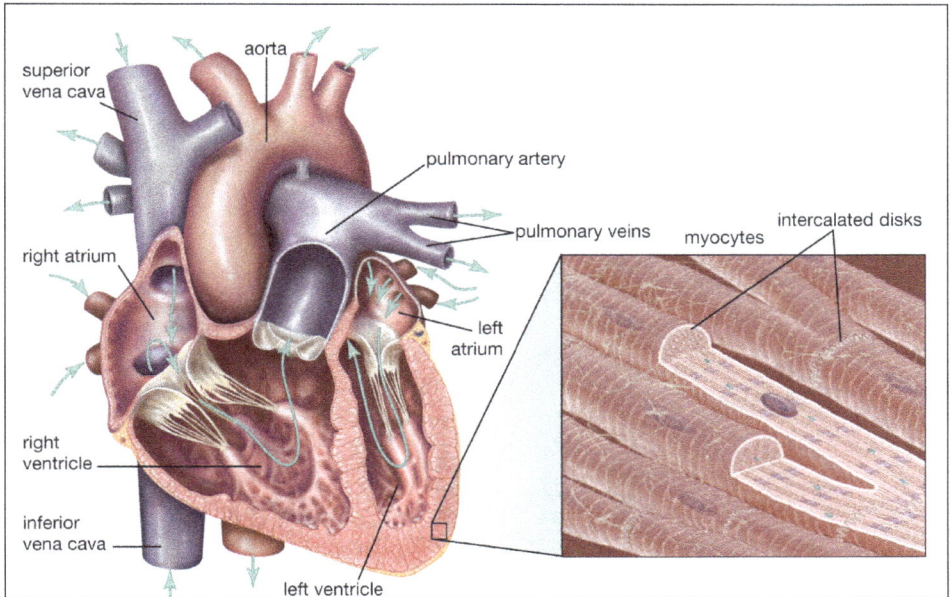

Mammalian heart cross section of a four-chambered mammalian heart.

Among vertebrates contraction of the heart is myogenic, or generated by muscle; rhythm is inherent in all cardiac muscle, but in myogenic hearts the pacemaker is derived from cardiac tissue.

The pacemaker in mammals (and also in birds) is an oblong mass of specialized cells called the sinoatrial node, located in the right atrium near the junction with the venae cavae. A wave of excitation spreads from this node to the atrioventricular node, which is located in the right atrium near the base of the interatrial septum. From this point excitation is conducted along the atrioventricular bundle (bundle of His) and enters the main mass of cardiac tissue along fine branches, the Purkinje fibres. Homeostatic, or stable, control of the heart by neuroendocrine or other agents is mediated through the intrinsic control network of the heart.

Blood leaves the left ventricle through the aorta. The mammalian aorta is an unpaired structure derived from the left fourth aortic arch of the primitive vertebrate. Birds, on the other hand, retain the right fourth arch.

The circulatory system forms a complex communication and distribution network to all physiologically active tissues of the body. A constant, copious supply of oxygen is required for sustaining the active, heat-producing (endothermous) physiology of the higher vertebrates. The efficiency of the four-chambered heart is important to this function. Oxygen is transported by specialized red blood cells, or erythrocytes, as in all vertebrates. Packaging the oxygen-bearing pigment hemoglobin in erythrocytes keeps the viscosity of the blood minimal and thereby allows efficient circulation while limiting the mechanical load on the heart. The mammalian erythrocyte is a highly evolved structure; its discoid, biconcave shape allows maximal surface area per unit volume. When mature and functional, mammalian red blood cells are enucleate (lacking a nucleus).

Respiratory System

Closely coupled with the circulatory system is the ventilatory (breathing) apparatus, the lungs and associated structures. Ventilation in mammals is unique. The lungs themselves are less efficient than those of birds, for air movement consists of an ebb and flow, rather than a one-way circuit, so a residual volume of air always remains that cannot be expired. Ventilation in mammals is by means of a negative pressure pump made possible by the evolution of a definitive thoracic cavity with a diaphragm.

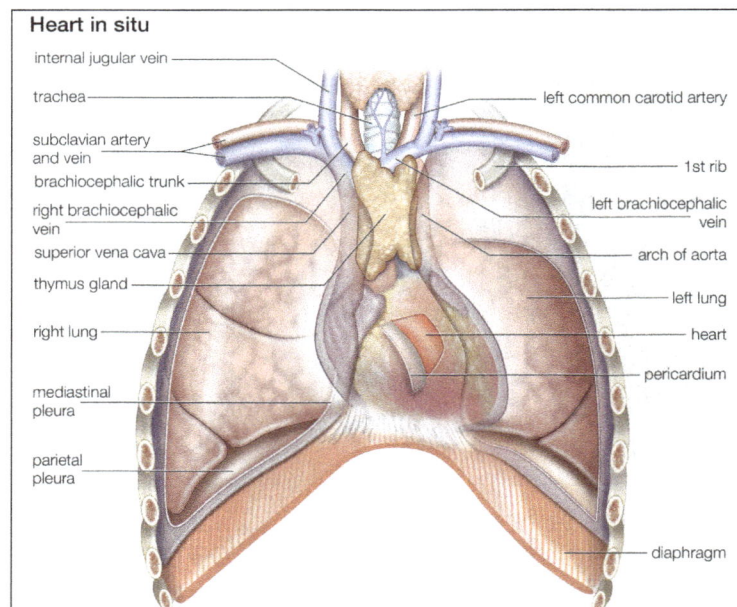

Heart in situ

internal jugular vein
trachea
subclavian artery and vein
brachiocephalic trunk
right brachiocephalic vein
superior vena cava
thymus gland
right lung
mediastinal pleura
parietal pleura

left common carotid artery
1st rib
left brachiocephalic vein
arch of aorta
left lung
heart
pericardium
diaphragm

Heart the human heart in situ.

The diaphragm is a unique composite structure consisting of (1) the transverse septum (a wall that primitively separates the heart from the general viscera); (2) pleuroperitoneal folds from the body wall; (3) mesenteric folds; and (4) axial muscles inserting on a central tendon, or diaphragmatic aponeurosis.

The lungs lie in separate airtight compartments called pleural cavities, separated by the mediastinum. As the size of the pleural cavity is increased, the lung is expanded and air flows in passively. Enlargement of the pleural cavity is produced by contraction of the diaphragm or by elevation of the ribs. The relaxed diaphragm domes upward, but when contracted it stretches flat. Expiration is an active movement brought about by contraction of abdominal muscles against the viscera.

Air typically enters the respiratory passages through the nostrils, where it may be warmed and moistened. It passes above the bony palate and the soft palate and enters the pharynx. In the pharynx the passages for air and food cross. Air enters the trachea, which divides at the level of the lungs into primary bronchi. A characteristic feature of the trachea of many mammals is the larynx. Vocal cords stretch across the larynx and are vibrated by forced expiration to produce sound. The laryngeal apparatus may be greatly modified for the production of complex vocalizations. In some groups—for example, howler monkeys—the hyoid apparatus is incorporated into the sound-producing organ, as a resonating chamber.

Nervous and Endocrine Systems

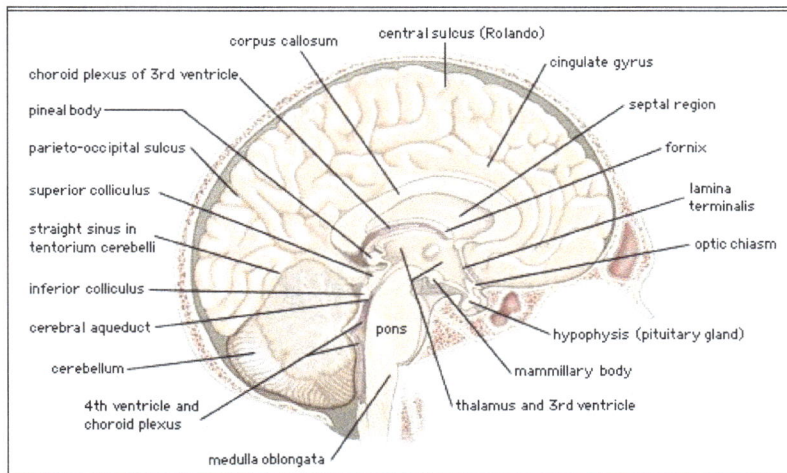

Medial view of the left hemisphere of the human brain.

The nervous system and the endocrine system are closely related to one another in their function, for both serve to coordinate activity. The endocrine glands of mammals generally have more complex regulatory functions than do those of lower vertebrates. This is particularly true of the pituitary gland, which supplies hormones that regulate the reproductive cycle. Follicle-stimulating hormone (FSH) initiates the maturation of the ovarian follicle. Luteinizing hormone (LH) mediates the formation of the corpus luteum from the follicle following ovulation. Prolactin, also a product of the anterior pituitary, stimulates the secretion of milk.

Control of the pituitary glands is partially by means of neurohumours from the hypothalamus, a part of the forebrain in contact with the pituitary gland by nervous and circulatory pathways.

The hypothalamus is of the utmost importance in mammals, for it integrates stimuli from both internal and external environments, channeling signals to higher centres or into autonomic pathways.

The cerebellum of vertebrates is at the anterior end of the hindbrain. Its function is to coordinate motor activities and to maintain posture. In most mammals the cerebellum is highly developed, and its surface may be convoluted to increase its area. The data with which the cerebellum works arrive from proprioceptors ("self-sensors") in the muscles and from the membranous labyrinth of the inner ear, the latter giving information on position and movements of the head.

In the vertebrate ancestors of mammals, the cerebral hemispheres were centres for the reception of olfactory stimuli. Vertebrate evolution has favoured an increasing importance of these lobes in the integration of stimuli. Their great development in mammals as centres of association is responsible for the "creative" behaviour of members of the class—i.e., the ability to learn, to adapt as individuals to short-term environmental change through appropriate responses on the basis of previous experience. In vertebrate evolution the gray matter of the cerebrum has moved from a primitive internal position in the hemispheres to a superficial position. The superficial gray matter is termed the pallium. The paleopallium of amphibians has become the olfactory lobes of the higher vertebrates; the dorsolateral surface, or archipallium, has become the mammalian hippocampus. The great neural advance of the mammals lies in the elaboration of the neopallium, which makes up the bulk of the cerebrum. The neopallium is an association centre, the dominant centre of neural function, and is involved in so-called "intelligent" response. By contrast, the highest centre in the avian brain is the corpus striatum, an evolutionary product of the basal nuclei of the amphibian brain. Therefore, the bulk of complex behaviour of birds is instinctive. The surface of the neopallium tends in some mammals to be greatly expanded by convoluting, forming folds (gyri) between deep grooves (sulci).

Primate Anatomy and Physiology

Head

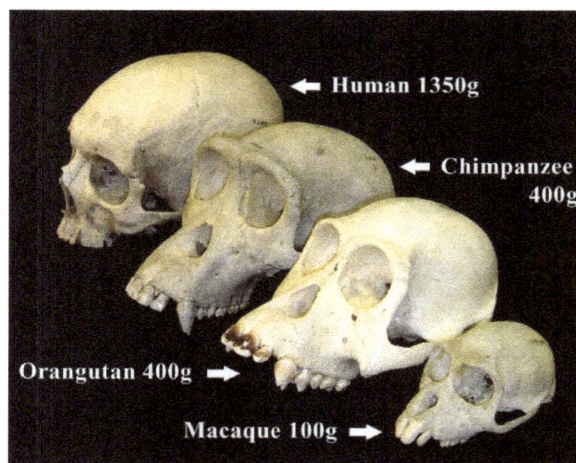

Primate skulls showing postorbital bar, and increasing brain sizes.

The primate skull has a large, domed cranium, which is particularly prominent in anthropoids. The cranium protects the large brain, a distinguishing characteristic of this group. The endocranial volume (the volume within the skull) is three times greater in humans than in the greatest nonhuman primate, reflecting a larger brain size. The mean endocranial volume is 1,201 cubic centimeters in humans, 469 cm³ in gorillas, 400 cm³ in chimpanzees and 397 cm³ in orangutans. The primary evolutionary trend of primates has been the elaboration of the brain, in particular the neocortex (a part of the cerebral cortex), which is involved with sensory perception, generation of motor commands, spatial reasoning, conscious thought and, in humans, language. While other mammals rely heavily on their sense of smell, the arboreal life of primates has led to a tactile, visually dominant sensory system, a reduction in the olfactory region of the brain and increasingly complex social behavior.

Primates have forward-facing eyes on the front of the skull; binocular vision allows accurate distance perception, useful for the brachiating ancestors of all great apes. A bony ridge above the eye sockets reinforces weaker bones in the face, which are put under strain during chewing. Strepsirrhines have a postorbital bar, a bone around the eye socket, to protect their eyes; in contrast, the higher primates, haplorhines, have evolved fully enclosed sockets.

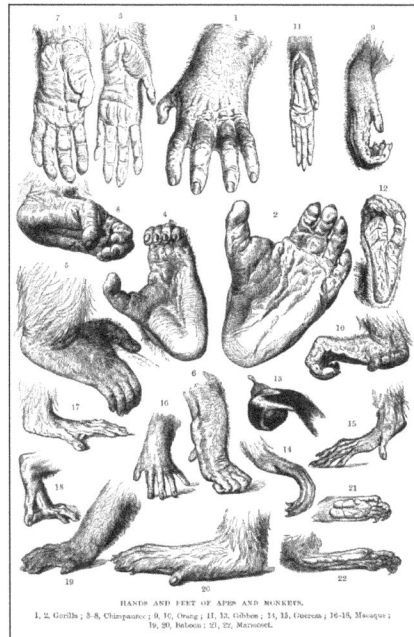

HANDS AND FEET OF APES AND MONKEYS.

1, 2, Gorilla ; 3-8, Chimpanzee ; 9, 10, Orang ; 11, 12, Gibbon ; 14, 15, Guereza ; 16-18, Macaque ; 19, 20, Baboon ; 21, 22, Marmoset.

Drawing of the hands and feet of various primates.

Primates show an evolutionary trend towards a reduced snout. Technically, Old World monkeys are distinguished from New World monkeys by the structure of the nose, and from apes by the arrangement of their teeth. In New World monkeys, the nostrils face sideways; in Old World monkeys, they face downwards. Dental pattern in primates vary considerably; although some have lost most of their incisors, all retain at least one lower incisor. In most strepsirrhines, the lower incisors form a toothcomb, which is used in grooming and sometimes foraging. Old World monkeys have eight premolars, compared with 12 in New World monkeys. The Old World species are divided into apes and monkeys depending on the number of cusps on their molars: monkeys have four, apes have five - although humans may have four or five. The main hominid molar cusp (hypocone) evolved in early primate history, while the cusp of the corresponding primitive lower molar (paraconid) was lost. Prosimians are distinguished by their immobilized upper lips, the moist tip of their noses and forward-facing lower front teeth.

Body

Vervet hindfoot showing fingerprint ridges on the sole.

Primates generally have five digits on each limb (pentadactyly), with keratin nails on the end of each finger and toe. The bottom sides of the hands and feet have sensitive pads on the fingertips. Most have opposable thumbs, a characteristic primate feature most developed in humans, though not limited to this order, (opossums and koalas, for example, also have them). Thumbs allow some species to use tools. In primates, the combination of opposing thumbs, short fingernails (rather than claws) and long, inward-closing fingers is a relict of the ancestral practice of gripping branches, and has, in part, allowed some species to develop brachiation (swinging by the arms from tree limb to tree limb) as a significant means of locomotion. Prosimians have clawlike nails on the second toe of each foot, called toilet-claws, which they use for grooming.

The primate collar bone is a prominent element of the pectoral girdle; this allows the shoulder joint broad mobility. Compared to Old World monkeys, apes have more mobile shoulder joints and arms due to the dorsal position of the scapula, broad ribcages that are flatter front-to-back, a shorter, less mobile spine, and with lower vertebrae greatly reduced - resulting in tail loss in some species. Prehensile tails are found in atelids, including the howler, spider, woolly spider, woolly monkeys; and in capuchins. Male primates have a pendulous penis and scrotal testes.

Locomotion

Primate species move by brachiation, bipedalism, leaping, arboreal and terrestrial quadrupedalism, climbing, knuckle-walking or by a combination of these methods. Several prosimians are primarily vertical clingers and leapers. These include many bushbabies, all indriids (i.e., sifakas, avahis and indris), sportive lemurs, and all tarsiers. Other prosimians are arboreal quadrupeds and climbers. Some are also terrestrial quadrupeds, while some are leapers. Most monkeys are both arboreal and terrestrial quadrupeds and climbers. Gibbons, muriquis and spider monkeys all brachiate extensively, with gibbons sometimes doing so in remarkably acrobatic fashion. Woolly monkeys also brachiate at times. Orangutans use a similar form of locomotion called quadramanous climbing, in which they use their arms and legs to carry their heavy bodies through the trees. Chimpanzees and gorillas knuckle walk, and can move bipedally for short distances. Although numerous species, such as australopithecines and early hominids, have exhibited fully bipedal locomotion, humans are the only extant species with this trait.

Diademed sifaka, a lemur that is a vertical clinger and leaper.

Vision

The *tapetum lucidum* of a northern greater galago, typical of prosimians, reflects the light of the photographers flash.

The evolution of color vision in primates is unique among most eutherian mammals. While the remote vertebrate ancestors of the primates possessed three color vision (trichromaticism), the nocturnal, warm-blooded, mammalian ancestors lost one of three cones in the retina during the Mesozoic era. Fish, reptiles and birds are therefore trichromatic or tetrachromatic, while all mammals, with the exception of some primates and marsupials, are dichromats or monochromats (totally color blind). Nocturnal primates, such as the night monkeys and bush babies, are often monochromatic. Catarrhines are routinely trichromatic due to a gene duplication of the red-green opsin gene at the base of their lineage, 30 to 40 million years ago. Platyrrhines, on the other hand, are trichromatic in a few cases only. Specifically, individual females must be heterozygous for two alleles of the opsin gene (red and green) located on the same locus of the X chromosome. Males, therefore, can only be dichromatic, while females can be either dichromatic or trichromatic. Color vision in strepsirrhines is not as well understood; however, research indicates a range of color vision similar to that found in platyrrhines.

Like catarrhines, howler monkeys (a family of platyrrhines) show routine trichromatism that has been traced to an evolutionarily recent gene duplication. Howler monkeys are one of the most specialized leaf-eaters of the New World monkeys; fruits are not a major part of their diets, and the type of leaves they prefer to consume (young, nutritive, and digestible) are detectable only by a red-green signal. Field work exploring the dietary preferences of howler monkeys suggests that routine trichromaticism was selected by environment.

Sexual Dimorphism in Non-human Primates

Sexual dimorphism describes the morphological, physiological, and behavioral differences between males and females of the same species. Most primates are sexually dimorphic for different biological characteristics, such as body size, canine tooth size, craniofacial structure, skeletal dimensions, pelage color and markings, and vocalization. However, such sex differences are primarily limited to the anthropoid primates; most of the strepsirrhine primates (lemurs and lorises) and tarsiers are monomorphic.

Hamadryas baboon female (left) and male (right).

Black howler monkey female (left) and male (right).

Types

Body Size

Extant primates exhibit a broad range of variation in sexual size dimorphism (SSD), or sexual divergence in body size. It ranges from species such as gibbons and strepsirrhines (including Madagascar's lemurs) in which males and females have almost the same body sizes to species such as chimpanzees and bonobos in which males' body sizes are larger than females' body sizes. In extreme cases, males have body sizes that are almost twice as large as those of females, as in some species including gorillas, orangutans, mandrills, hamadryas baboons, and proboscis monkeys. Patterns of size dimorphism exhibited in primates may correspond to the intensity of competition between members of the same sex for access to mates—intrasexual competition, counteracted by fecundity selection on the other sex. Some callitrichine and strepsirrhine primates are, however, characterized by the reverse dimorphism, a phenomenon in which females are larger than males. For lemurs, for example, females' dominance over males accounts for the reverse dimorphism.

Tooth Size

Canine sexual dimorphism is one particular type of sexual dimorphism, in which males of a species have larger canines than females. Within primates, the male and female canine tooth size varies among different taxonomic subgroups, yet canine dimorphism is most extensively found in catarrhines among haplorhine primates. For example, in many baboons and macaques, the size of male canines is more than twice as large as that of female canines. It is rare, yet females in some species are known to have larger canines than males, such as the eastern brown mouse lemur (Microcebus rufus). Sexual dimorphism in canine tooth size is relatively weak or absent in extant strepsirrhine primates. The South American titi monkeys (Callicebus moloch), for instance, do not exhibit any differences in the size of canine teeth between the sexes.

Among different types of teeth constituting the dentition of primates, canines exhibit the greatest degree of variation in tooth size, whereas incisors have less variation and cheek teeth have the least. A canine dimorphism is also more widely seen in maxillary canines than in mandibular canines.

Craniofacial Structure

Craniofacial sex differentiation among anthropoid primates varies in a wide range and is known to arise primarily through ontogenetic processes. Studies on hominids have shown that, in general, males tend to have a greater increase of facial volume than of neurocranial volume, a more obliquely oriented foramen magnum, and a more pronounced rearrangement of the nuchal region. The breadth, length and height of the neurocranium in adult male macaques, guenons, orangutans and gorillas are about nine percent larger than the neurocranial dimensions in adult females, whereas in spider monkeys and gibbons the sex differences is on a general average about 4 to 5 percent. In orangutans, males and females share similarities in facial dimensions and growth in terms of orbits, nasal width, and facial width. They tend to have some significant differences, however, in various facial heights (e.g., height of the anterior face, premaxilla, and nose).

Skeletal Structure

Primates also exhibit sexual dimorphism in skeletal structures. In general, skeletal dimorphism in primates is primarily known as a product of body mass dimorphism. Hence, males have proportionally larger skeletons compared to females due to their larger body masses. Larger and more robust skeletal structures in males is also attributable to better developed muscle scarring, and more intense cresting of bones compared to those of females. Male gorillas, for example, possess large sagittal and nuchal crests, which correspond to their large temporalis muscles and nuchal musculature. Also, an unusual skeletal dimorphism includes enlarged, hollow hyoid bones found in males of gibbons and howler monkeys, which contribute to the resonation of their voices.

Pelage Color and Markings

Sex differences in pelage, such as capes of hair, beards, or crests, and skin can be found in several species among adult primates. Several species (e.g., *Lemur macaco*, *Pithecia pithecia*, *Alouatta caraya*) show an extensive dimorphism in pelage colors or patterning. For example, in mandrills (*Mandrillus sphinx*), males display extensive red and blue coloration on their face, rump and genitalia as compared to females. Male mandrills also possess a yellow beard, nuchal crest of hair, and pronounced boney paranasal ridges, all of which are absent or vestigial in females. Studies have shown that male color in mandrills serves as a badge of social status in the species.

Temporary Sexual Dimorphism

Some sexual dimorphic traits in primates are known to appear on a temporary basis. In squirrel monkeys (*Saimiri sciureus*), males can gain fat as much as 25 percent of the body mass only during the breeding season, specifically in their upper torso, arms, and shoulders. This seasonal phenomenon, known as "male fattening," is associated with both male-male competition and female choice for larger males. Orangutan males tend to gain weight and develop large cheek flanges, when they achieve dominance over other group members.

Vocalization

In many adult primates, dimorphism in the vocal repertoire can appear in both call production (e.g., calls with a particular set of acoustic traits) and usage (e.g., call frequency and context-specificity) between the sexes. Sex-specific calls are commonly found in Old World monkeys, in which males produce loud calls for intergroup spacing and females produce copulation calls for sexual activity. Forest guenons also tend to display strong vocal divergences between sexes, with mostly sex-specific call types. Studies on De Brazza's monkeys (Cercopithecus neglectus), one of the African guenon species, have shown that call rates in adult females (24 call. hr-1) are more than seven times higher than in adult males (2.5call.hr-1). A usage of different call types also differs between sexes, in that females mostly utter contact(-food) calls, whereas males produce a great number of threat calls. Such difference in vocal usage is associated with social roles, with females being involved in more social tasks within the group and males being responsible for territory defense.

Ultimate Mechanisms

Ultimate mechanisms for sexual dimorphism in primates explain the evolutionary history and functional significance of the sexual dimorphism expressed among primates.

Sexual Selection

In primates, sexual dimorphism including body size, canine tooth size, and morphological characteristics is often attributed to sexual selection, which is believed to act through two mechanisms: intrasexual competition and female mate choice.

Most male anthropoid primates increase their potential reproductive output by directly engaging in agonistic (contest) competition for gaining access to females. Any weaponry or other physical characteristics that allow males to win intrasexual combat are therefore strongly favored for the selection. Larger body size has been thought to confer advantages to males in competition for access to females, which is consistent with sexual selection hypothesis. Males with a larger canine tooth also tend to be competitively superior to males with a smaller canine, which explain a dimorphism in canine size between the sexes. For example, baboons are highly dimorphic in both body mass and canine size, where males are actively engaged in fights for increasing their mating success and defending females against other males.

Differential parental investment between the sexes accounts for female mate choice. The number of offspring produced by female primates is often limited due to the small litter size, long intervals between births, relatively slow-growing offspring, and energetically expensive costs of pregnancy, lactation, and child care. Females thus choose their mates possessing certain preferable traits, which could possibly provide genetic or direct phenotypic benefits. For example, the large mane found in male gelada (*Theropithecus*) is assumed to be a preferable pelage condition favored by females, who primarily control and select their mates. Such preference leads the increase in size dimorphism across primate species, which may be favorable in an environment where resources are limited.

Mating System

A strong association between polygynous mating system and dimorphism in primates has been observed. Monogamous species tend to show lower degree of sexual dimorphism than polygynous species, since monogamous males have a lower differential reproductive success. Monogamous mating system seems to account for minimal dimorphism in hylobatids, in which females are codominant with males. As an exception, among polygynous primates, colobines as a group consistently exhibit a low level of sexual size dimorphism for unclear reasons. In terms of canine dimorphism, males in polygynous species tend to have larger and relatively stronger canines than males in monogamous and polyandrous species.

Phylogeny

Similar magnitudes of body weight dimorphism have been observed in all species within several taxonomic groups such as callitrichids, hylobatids, *Cercopithecus*, and *Macaca*. Such correlation between phylogenetic relatedness and sexual dimorphism across different groups reflects similarities in their behaviors and ecological conditions, but not in independent adaptations. This idea is referred to as "phylogenetic niche conservatism."

Terrestriality

Terrestrial primates tend to show a greater degree of dimorphism than arboreal primates. It has been hypothesized that larger sizes of body mass and canine tooth are favored among males of terrestrial primates due to the likelihood of higher vulnerability to predators. Another hypothesis suggests that arboreal primates have limitations on their upper body size, given that larger body size could disrupt their usage of terminal branches for locomotion. However, among some species of guenons (*Cercopithecus*), arboreal blue monkeys (*C. mitis*) appear to be more sexually dimorphic than terrestrial vervet monkeys (*C. aethiops*).

Niche Divergence

It has been hypothesized that niche divergence between the sexes attributes to the evolution of size dimorphism in primates. Males and females are known to have different preferences for ecological habitat due to different reproductive activities, which could possibly lead to dietary differences, followed by dimorphic morphological traits. This niche divergence hypothesis, however, has never been strongly supported due to the lack of compelling data.

Toothcomb

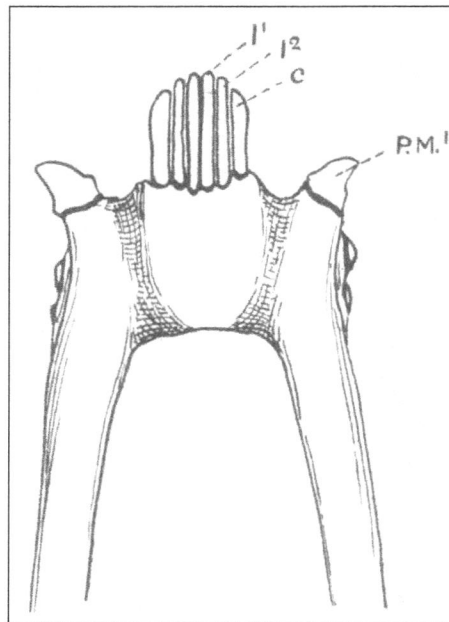

The lemuriform toothcomb, viewed from the underside of the lower jaw.

A toothcomb (also tooth comb or dental comb) is a dental structure found in some mammals, comprising a group of front teeth arranged in a manner that facilitates grooming, similar to a hair comb. The toothcomb occurs in lemuriform primates (which includes lemurs and lorisoids), treeshrews, colugos, hyraxes, and some African antelopes. The structures evolved independently in different types of mammals through convergent evolution and vary both in dental composition and structure. In most mammals the comb is formed by a group of teeth with fine spaces between them. The toothcombs in most mammals include incisors only, while in lemuriform primates they include incisors and canine teeth that tilt forward at the front of the lower jaw, followed by a canine-shaped first premolar. The toothcombs of colugos and hyraxes take a different form with the individual incisors being serrated, providing multiple tines per tooth.

The toothcomb is usually used for grooming. While licking the fur clean, the animal will run the toothcomb through the fur to comb it. Fine grooves or striations are usually cut into the teeth during grooming by the hair and may be seen on the sides of the teeth when viewed through a scanning electron microscope. The toothcomb is kept clean by either the tongue or, in the case of lemuriforms, the sublingua, a specialized "under-tongue". The toothcomb can have other functions, such as food procurement and bark gouging. Within lemuriforms, fork-marked lemurs and indriids have more robust toothcombs to support these secondary functions. In some lemurs, such as the aye-aye, the toothcomb has been lost completely and replaced with other specialized dentition.

In lemuriform primates, the toothcomb has been used by scientists in the interpretation of the evolution of lemurs and their kin. They are thought to have evolved from early adapiform primates around the Eocene or earlier. One popular hypothesis is that they evolved from European adapids, but the fossil record suggests that they evolved from an older lineage that migrated to Africa during the Paleocene (66 to 55 mya) and might have evolved from early cercamoniines from Asia. Fossil primates such as *Djebelemur*, *'Anchomomys' milleri*, and *Plesiopithecus* may have been their closest relatives. The lack of a distinct toothcomb in the fossil record before to 40 mya has created a conflict with molecular clock studies that suggest an older divergence between lemurs and lorisoids, and the existence of a ghost lineage of lemuriform primates in Africa.

Homologous and Analogous Structures

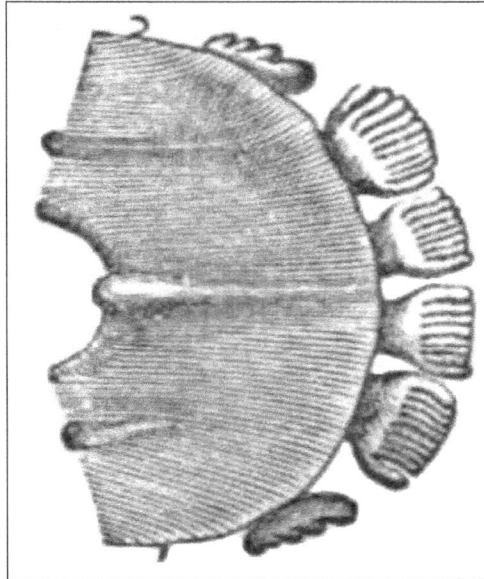

The toothcomb of a colugo differs significantly in shape from that of lemuriform primates.

The toothcomb, a special morphological arrangement of teeth in the anterior lower jaw, is best known in extant strepsirrhine primates, which include lemurs and lorisoid primates (collectively known as lemuriforms[a]). This homologous structure is a diagnostic character that helps define this clade (related group) of primates. An analogous trait is found in the bald uakari (*Cacajao calvus*), a type of New World monkey.

Toothcombs can also be found in colugos and treeshrews, both close relatives of primates; however, the structures are different and these are considered to examples of convergent evolution. Likewise, small- or medium-sized African antelopes, such as the impala (*Aepyceros melampus*), have a similar structure sometimes referred to as the "lateral dental grooming apparatus". Living and extinct hyraxes (hyracoids) also exhibit a toothcomb, although the number of tines in the comb vary throughout the fossil record.

Dating to the Eocene epoch over 50 mya, *Chriacus* and *Thryptacodon*—two types of arctocyonids (primitive placental mammals)—also possessed an independently evolved toothcomb.

Anatomical Structure

The toothcomb of most lemuriforms includes six finely spaced teeth, four incisors and two canine teeth that are procumbent (tilt forward) in the front of the mouth. The procumbent lower canine teeth are the same shape as the incisors located between them, but they are more robust and curve upward and inward, more so than the incisors. In the permanent dentition, the canines erupt after the incisors. The crowns of the incisors are also angled in the direction of the forward tilt, and the crowns of both the incisors and canines are elongated and compressed side-to-side. The apical ridge, following along the front edges of the toothcomb teeth, is V-shaped in most lemuriforms, tapering off from the midline. As a result of this dental reconfiguration, the upper and lower incisors do not contact one another, and often the upper incisors are reduced or lost completely.

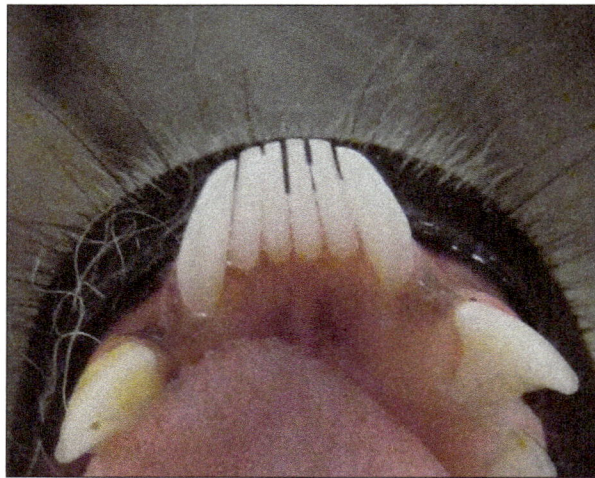

Toothcomb of a ring-tailed lemur, with canine-like premolars behind it.

The French anatomist Henri Marie Ducrotay de Blainville first identified the two lateral teeth of the lemuriform toothcomb as canines in 1840. Canine teeth are normally used to pierce or grasp objects. With modified lower canine teeth, the first lower premolars following the toothcomb are usually shaped like typical canine teeth (caniniform) and assume their function. These premolars are commonly confused with canines. Normally the true canines in the lower jaw sit in front of the upper canines, and in toothcombed primates, the caniniform premolars rest behind it.

The lemuriform toothcomb is kept clean by the sublingua or "under-tongue", a specialized muscular structure that acts like a toothbrush to remove hair and other debris. The sublingua can extend below the end of the tongue and is tipped with keratinized, serrated points that rake between the front teeth.

Among lemurs, the toothcomb is variable in structure. Among indriids (Indriidae), the toothcomb is less procumbent and consists of four teeth instead of six. The indriid toothcomb is more robust and wider, with shorter incisors, wider spaces between the teeth (interdental spaces), and a broader apical ridge. It is unclear whether this four-toothed toothcomb consists of two pairs of incisors or one pair of incisors and one pair of canines. In fork-marked lemurs (Phaner) the toothcomb is more compressed, with significantly reduced interdental spaces. All six teeth are longer, straighter, and form a more continuous apical ridge. In the recently extinct monkey lemurs (Archaeolemuridae) and sloth lemurs (Palaeopropithecidae), the toothcomb was lost and the incisors and canines resumed a typical configuration in the front of the mouth. The aye-aye also lost its toothcomb, replacing it with continually growing (hypselodont) front teeth, similar to the incisors of rodents.

In colugos, the toothcomb has a completely different structure. Instead of individual incisors and canine teeth being finely spaced to act like the teeth of a comb, the biting edge of the four incisors have become serrated with as many as 15 tines each, while the canine acts more like a molar. These serrated incisors are kept clean using the front of the tongue, which is serrated to match the serrations of the incisors. Similarly, the hyracoid toothcomb consists of incisors with multiple tines, called "pectinations". In contrast to the colugos, the size and shape of the tines are more uniform.

The toothcomb of treeshrews is like the lemuriform toothcomb in that it uses interdental spaces to form the comb tines, but only two of its three pairs of lower incisors are included in the toothcomb and the canines are also excluded. The lateral two incisors in the toothcomb are generally larger. In the extinct arctocyonids, all six lower incisors were part of the toothcomb. In African antelopes, the toothcomb is strikingly similar to that of lemuriforms in that it consists of two pairs of incisors and a pair of canines.

Functions

As a homologous structure in lemuriforms, the toothcomb serves variable biological roles, despite its superficially stereotypic shape and appearance. It is primarily used as a toiletry device or grooming comb. Additionally, some species use their toothcomb for food procurement or to gouge tree bark.

Grooming

The sublingua is a secondary tongue below the primary tongue and is used to
remove hair and debris from the toothcomb of lemuriforms.

The primary function of the toothcomb, grooming, was first noted by the French naturalist Georges Cuvier in 1829, who pointed out that the ring-tailed lemur (*Lemur catta*) had lower incisors that "*sont de véritables peignes*" ("are real combs"). More than 100 years later, the grooming function was questioned since it was difficult to observe and the interdental spaces were thought to be too small for fur. Observations later showed the teeth were used for that purpose and that immediately after grooming, hair may be found trapped in the teeth, but is removed by the sublingua later.

In 1981, scanning electron microscopy revealed fine grooves or striations on the teeth in lemuriform toothcombs. These grooves were only found on the sides of the teeth on the concave surfaces between the sides, as well on the back ridge of the teeth. Between 10 and 20 µm wide, these grooves indicate that hair moved repeatedly across the teeth. Inside these grooves were even finer grooves, less than 1 µm, created by abrasion with the cuticular layer of the hair.

Among non-primates, the extinct *Chriacus* exhibits microscopic groves on its toothcomb, but the Philippine colugo (*Cynocephalus volans*) does not. The toothcomb of the colugos is generally considered to function as a toothcomb, but due to the lack of striations on the teeth and no documented observations of toothcomb use during oral grooming, its use seems to be limited to food procurement.

In African antelopes, the lateral dental grooming apparatus does not appear to be used during grazing or browsing. Instead, it is used during grooming when the head sweeps upward in a distinctive motion. It is thought to comb the fur and remove ectoparasites.

Olfaction in Lemuriforms

In lemuriform primates, the toothcomb may also play a secondary role in olfaction, which may account for the size reduction of the poorly studied upper incisors. The toothcomb may provide pressure to stimulate glandular secretions which are then spread through the fur. Furthermore, the size reduction of the upper incisors may create a gap between the teeth (interincisal diastema) that connects the philtrum (a cleft in the middle of the wet nose, or rhinarium) to the vomeronasal organ in the roof of the mouth. This would allow pheromones to be more easily transferred to the vomeronasal organ.

Food Procurement and other Uses

Mouse lemurs (*Microcebus*), sifakas (*Propithecus*), and the indri (*Indri*) use their toothcombs to scoop up fruit pulp. Other small lemuriforms, such as fork-marked lemurs (*Phaner*), the hairy-eared dwarf lemur (*Allocebus*), and galagos (particularly the genera *Galago* and *Euoticus*) use their toothcombs to tooth-scrape plant exudates, such as gum and sap. In fork-marked lemurs, the toothcomb is specially adapted to minimize food trapment since the interdental spaces are greatly reduced. The herbivorous colugos in the genus *Cynocephalus* may also use their toothcomb for food procurement.

Indriids such as the sifakas use their toothcombs to gouge bark or dead wood (bark-prising), which is done before scent-marking with the gland on their chest. The more robust structure of their toothcomb is thought to help it withstand the compressive forces experienced during regular bark-prising.

Primate Basal Ganglia

The basal ganglia form a major brain system in all species of vertebrates, but in primates (including humans) there are special features that justify a separate consideration. As in other vertebrates, the primate basal ganglia can be divided into striatal, pallidal, nigral, and subthalamic components. In primates, however, there are two pallidal subdivisions called the external globus pallidus (GPe) and internal globus pallidus (GPi). Also in primates, the dorsal striatum is divided by a large tract called the internal capsule into two masses named the caudate nucleus and the putamen—in most other species no such division exists, and only the striatum as a whole is recognized. Beyond this, there is a complex circuitry of connections between the striatum and cortex that is specific to primates. This complexity reflects the difference in functioning of different cortical areas in the primate brain.

Functional imaging studies have been performed mainly using human subjects. Also, several major degenerative diseases of the basal ganglia, including Parkinson's disease and Huntington's disease, are specific to humans, although "models" of them have been proposed for other species.

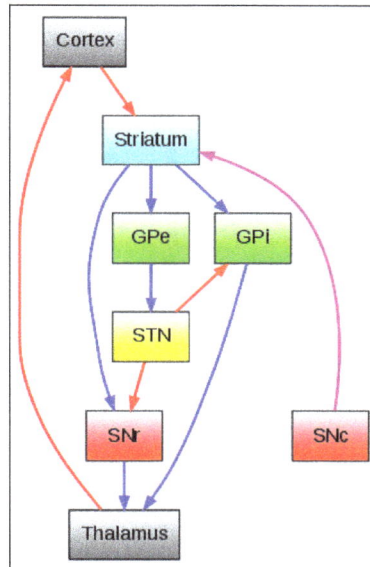

Diagram of the main components of the basal ganglia and their interconnections.

Corticostriatal Connection

A major output from the cortex, with axons from most of the cortical regions connecting to the striatum, is called the corticostriatal connection, part of the cortico-basal ganglia-thalamo-cortical loop. In the primate most of these axons are thin and unbranched. The striatum does not receive axons from the primary olfactory, visual or auditory cortices. The corticostriatal connection is an excitatory glutamatergic pathway. One small cortical site can project many axon branches to several parts of the striatum.

Striatum

The striatum is the largest structure of the basal ganglia.

Structure

Neuronal Constitution

Medium spiny neurons (MSN)s, account for up to 95 per cent of the striatal neurons. There are two populations of these projection neurons, MSN1 and MSN2, both of which are inhibitory GABAergic. There are also various groups of GABAergic interneurons and a single group of cholinergic interneurons. These few types are responsible for the reception, processing, and relaying of all the cortical input.

Most of the dendritic spines on the medium spiny neurons synapse with cortical afferents and their axons project numerous collaterals to other neurons. The cholinergic interneurons of the primate, are very different from those of non-primates. These are said to be tonically active.

The dorsal striatum and the ventral striatum have different populations of the cholinergic interneurons showing a marked difference in shape.

Physiology

Unless stimulated by cortical input the striatal neurons are usually inactive.

Levels of Organisation

The striatum is one mass of grey matter that has two different parts, a ventral and a dorsal part. The dorsal striatum contains the caudate nucleus and the putamen, and the ventral striatum contains the nucleus accumbens and the olfactory tubercle. The internal capsule is seen as dividing the two parts of the dorsal striatum. Sensorimotor input is mostly to the putamen. An associative input goes to the caudate nucleus and possibly to the nucleus accumbens.

There are two different components of the striatum differentiated by staining – striosomes and a matrix. Striosomes are located in the matrix of the striatum and these contain μ-opioid receptors and dopamine receptor D1 binding sites.

The striatopallidal fibers give a connection from the putamen to the globus pallidus and substantia nigra.

Connectomics

Unlike the inhibitory GABAergic neurons in the neocortex that only send local connections, in the striatum these neurons send long axons to targets in the pallidum and substantia nigra. A study in macaques showed that the medium spiny neurons have several targets. Most striatal axons first target the GPe, some of these also target the GPi and both parts of the substantia nigra. There are no single axon projections to either the GPi, or to the SN, or to both of these areas; only connecting as continuing targets via axon collaterals from the striatum to the GPe.

The only difference between the axonal connectomes of the striosomes and the axons of those neurons in the matrix, is in the numbers of their branching axons. Striosomal axons cross the extent of the SN, and in macaques emit 4 to 6 vertical collaterals that form vertical columns which enter deep into the SN pars compacta (SNpc); the axons from those in the matrix are more sparsely branched. This pattern of connectivity is problematic. The main mediator of the striatopallidonigral system is

GABA and there are also cotransmitters. The GPe stains for met-enkephalin, the GPi stains for either substance P or dynorphin or both, and the SN stains for both. This probably means that a single axon is able to concentrate different co-mediators in different subtrees, depending on the target.

Selectivity of Striatal Territories for Targets

A study of the percentage of striatal axons from the sensorimotor (putamen) and associative striatum (caudate nucleus) to the globus pallidus found important differences. The GPe for instance receives a large input of axons from the associative areas. The GPi is strongly sensorimotor connected. The SN is at first associative. This is confirmed by the effects of striatal stimulations.

All the projections from the primary somatosensory cortex to the putamen, avoid the striosomes and innervate areas within the matrix.

Pallidonigral Set and Pacemaker

Constitution

The pallidonigral set comprises the direct targets of the striatal axons: the two nuclei of the pallidum, and the pars compacta (SNpc) and pars reticulata (SNpr) of the substantia nigra. One character of this ensemble is given by the very dense striato-pallidonigral bundle giving it its whitish aspect (pallidus means pale). In no ways has the pallidum the shape of a globe. After Foix and Nicolesco and some others, Cécile and Oskar Vogt suggested the term pallidum - also used by the Terminologia Anatomica. They also proposed the term nigrum for replacing nigra, which is indeed not a substance; but this is generally not followed. The whole pallidonigral set is made up the same neuronal components. The majority is made up of very large neurons, poorly branched, strongly stained for parvalbumin, having very large dendritic arborisations (much larger in primates than in rodents) with straight and thick dendrites. Only the shape and direction of the dendritic arborizations differ between the pallidum and the SN neurons. The pallidal dendritic arborisations are very large flat and disc-shaped. Their principal plane is parallel to the others and also parallel to the lateral border of the pallidum; thus perpendicular to the axis of the afferences. Since the pallidal discs are thin, they are crossed only for a short distance by striatal axons. However, since they are wide, they are crossed by many striatal axons from wide striatal parts. Since they are loose, the chances of contact are not very high. Striatal arborisations, emit perpendicular branches participating in flat bands parallel to the lateral border, which increases the density of synapses in this direction. This is true for not only for the striatal afferent but also for the subthalamic. The synaptology of the set is uncommon and characteristic. The dendrites of the pallidal or nigral axons are entirely covered by synapses, without any apposition of glia. More than 90% of synapses are of striatal origin. One noticeable property of this ensemble is that not one of its elements receives cortical afferents. Initial collaterals are present. However, in addition to the presence of various appendages at the distal extremity of the pallidal neurons that could act as elements of local circuitry, there are weak or no functional interrelations between pallidal neurons.

External Globus Pallidus

The external globus pallidus (GPe) or lateral globus pallidus, is flat, curved and extended in depth and width. The branching dendritic trees are disc-shaped, flat, run parallel to each other and to the pallidum border, and are perpendicular to those axons coming from the striatum. The GPe also receives input from the subthalamic nucleus, and dopaminergic input from the SNpc. The GPe does not give output to the thalamus only intrasystemically connecting to the other basal ganglia structures. It can be seen as a GABA inhibitory mediator regulating the basal ganglia. Its firing activity is very fast and exhibits long intervals of up to several seconds of silence.

In monkeys an initial inhibition was seen in response to striatal input, followed by a regulated excitation. In the study this suggested that the excitation was used temporarily to control the magnitude of the incoming signal and to spatially focus this into a limited number of pallidal neurons. GPe neurons are often multi-targeted and may respond to a number of neuron types. In macaques, axons from the GPe to the striatum account for about 15%; those to the GPi, SNpr and subthalamic nucleus are about 84%. The subthalamic nucleus was seen to be the preferred target which also sends most of its axons to the GPe.

Internal Globus Pallidus

The internal globus pallidus (GPi) or medial globus pallidus is only found in the primate brain and so is a younger portion of the globus pallidus. Like the GPe and the substantia nigra the GPi is a fast-spiking pacemaker but its activity does not show the long intervals of silence seen in the others. In addition to the striatal input there is also dopaminergic input from the SNpc. Unlike the GPe the GPi does have a thalamic output and a smaller output towards the habenula. It also gives output to other areas including the pedunculopontine nucleus and to the area behind the red nucleus. The evolutionary increase of the internal pallidus also brought an associated increase in the pallidothalamic tracts, and the appearance of the ventral lateral nucleus in the thalamus. The mediator is GABA.

Substantia Nigra

The substantia nigra is made up of two parts, the pars compacta (SNpc) and the pars reticulata (SNpr), sometimes there is a reference to the pars lateralis but that is usually included as part of the pars reticulata. The "black substance" that the term translates as, refers to the neuromelanin found in the dopaminergic neurons. These are found in a darker region of the SNpc. The SNpr is a lighter coloured region. There are similar cells in the substantia nigra and the globus pallidus. Both parts receive input from the striatopallidal fibres.

Pars Compacta

The pars compacta is the most lateral part of the substantia nigra and sends axons to the superior colliculus. The neurons have high firing rates which make them a fast-spiking pacemaker and they are involved in ocular saccades.

Pars Reticulata

The border between the SNpc and SNpr is highly convoluted with deep fringes. Its neuronal genus is the same as that of the pallidum, with the same thick and long dendritic trees. It receives its synapses from the striatum in the same way as the pallidum. Striatonigral axons from the striosomes may form columns vertically oriented entering deeply in the SNpr. The ventral dendrites of the SNpc from the reverse direction go also deeply in it. The SN also send axons to the pedunculopontine nucleus. and to the parafascicular part of the central complex. The SNpr is another "fast-spiking pacemaker" Stimulations provoke no movements. Confirming anatomical data, few neurons respond to passive and active movements (there is no sensorimotor map) but a large proportion shows responses that may be related to memory, attention or movement preparation that would correspond to a more elaborate level than that of the medial pallidum. In addition to the massive striatopallidal connection, the SNpr receives a dopamine innervation from the SNpc and glutamatergic axons from the pars parafascicularis of the central complex. It sends nigro-thalamic axons. There is no conspicuous nigro-thalamic bundle. Axons arrive medially to the pallidal afferences at the anterior and most medial part of the lateral region of the thalamus: the ventral anterior nucleus (VA) differentiated from the ventral lateral nucleus (VL) receiving pallidal afferences. The mediator is GABA.

Striatopallidonigral Connection

The striatopallidonigral connection is a very particular one. It engages the totality of spiny striatal axons. Estimated numbers are 110 million in man, 40 in chimpanzees and 12 in macaques. The striato-pallido-nigral bundle is made up of thin, poorly myelinated axons from the striatal spiny neurons grouped into pencils "converging like the spokes of a wheel". It gives its "pale" aspect to the receiving areas. The bundle strongly stains for iron using Perls' Prussian blue (in addition to iron it contains many heavy metals including cobalt, copper, magnesium and lead).

Convergence and Focusing

After the huge reduction in number of neurons between the cortex and the striatum, the striatopallido-nigral connection is a further reduction in the number of transmitting compared to receiving neurons. Numbers indicate that, for 31 million striatal spiny neurons in macaques, there are only 166000 lateral pallidal neurons, 63000 medial pallidal, 18000 lateral nigral and 35000 in the pars reticulata. If the number of striatal neurons is divided by their total number, as an average, each target neuron may receive information from 117 striatal neurons. (Numbers in man lead to about the same ratio). A different approach starts from the mean surface of the pallidonigral target neurons and the number of synapses that they may receive. Each pallidonigral neuron may receive 70000 synapses. Each striatal neuron may contribute 680 synapses. This leads again to an approximation of 100 striatal neurons for one target neuron. This represents a huge, infrequent, reduction in neuronal connections. The consecutive compression of maps cannot preserve finely distributed maps (as in the case for instance of sensory systems). The fact that a strong anatomical possibility of convergence exists does not means that this is constantly used. A recent modeling study starting from entirely 3-d reconstructed pallidal neurons showed that their morphology alone is able to create a center-surround pattern of activity. Physiological analyses have shown a central inhibition/peripheral excitation pattern, able of focusing the pallidal response in normal conditions. Percheron and Filion thus argued for a "dynamically focused convergence". Disease, is able to alter the normal focusing. In monkeys intoxicated by MPTP, striatal stimulations lead to a large convergence on pallidal neurons and a less precise mapping. Focusing is not a property of the striatopallidal system. But, the very particular and contrasted geometry of the connection between striatal axons and pallidonigral dendrites offers particular conditions (the possibility for a very large number of combinations through local additions of simultaneous inputs to one tree or to several distant foci for instance). The disfocusing of the system is thought to be responsible for most of the parkinsonian series symptoms. The mechanism of focusing is not known yet. The structure of the dopaminergic innervation does not seem to allow it to operate for this function. More likely focusing is regulated by the upstream striatopallidal and corticostriatal systems.

Synaptology and Combinatory

The synaptology of the striato- pallidonigral connection is so peculiar as to be recognized easily. Pallidonigral dendrites are entirely covered with synapses without any apposition of glia. This gives in sections characteristic images of "pallissades" or of "rosettes". More than 90% of these synapses are of striatal origin. The few other synapses such as the dopaminergic or the cholinergic are interspersed among the GABAergic striatonigral synapses. The way striatal axons distribute their synapses is a disputed point. The fact that striatal axons are seen parallel to dendrites as "woolly fibers" has led to exaggerate the distances along which dendrites and axons are parallel. Striatal axons may in fact simply cross the dendrite and give a single synapse. More frequently the striatal axon curves its course and follow the dendrite forming "parallel contacts" for a rather short distance. The average length of parallel contacts was found to be 55 micrometres with 3 to 10 boutons (synapses). In another type of axonal pattern the afferent axon bifurcates and gives two or more branches, parallel to the dendrite, thus increasing the number of synapses given by one striatal axon. The same axon may reach other parts of the same dendritic arborisation (forming "random cascades") With this pattern, it is more than likely that 1 or even 5 striatal axons are not able to influence (to inhibit) the activity of one pallidal neuron. Certain spatio-temporal conditions would be necessary for this, implying more afferent axons.

Pallidonigral Outmaps

What is described above concerned the input map or "inmap" (corresponding to the spatial distribution of the afferent axons from one source to one target). This does not correspond necessarily to the output map or outmap (corresponding to the distribution of the neurons in relation to their axonal targets). Physiological studies and transsynaptic viral markers have shown that islands of pallidal neurons (only their cell bodies or somata, or trigger points) sending their axons through their particular thalamic territories (or nuclei) to one determined cortical target are organized into radial bands. These were assested to be totally representative of the pallidal organisation. This is certainly not the case. Pallidum is precisely one cerebral place where there is a dramatic change between one afferent geometry and a completely different efferent one. The inmap and the outmap are totally different. This is an indication of the fundamental role of the pallidonigral set: the spatial reorganisation of information for a particular "function", which is predictably a particular reorganisation within the thalamus preparing a distribution to the cortex. The outmap of the nigra (lateralis reticulata) is less differentiated.

Substantia Nigra Compacta (Snpc) and Nearby Dopaminergic Elements

In strict sense, the pars compacta is a part of the core of basal ganglia core since it directly receives synapses from striatal axons through the striatopallidonigral bundle. The long ventral dendrites of the pars compacta indeed plunge deep in the pars reticulata where they receive synapses from the bundle. However, its constitution, physiology and mediator contrast with the rest of the nigra. This explains why it is analysed here between the elements of the core and the regulators. Ageing leads to the blackening of its cell bodies, by deposit of melanin, visible by naked eye. This is the origin of the name of the ensemble, first "locus niger" (Vicq d'Azyr), meaning black place, and then "substantia nigra" (Sömmerring), meaning black substance.

Structure

The densely distributed neurons of the pars compacta have larger and thicker dendritic arborizations than those of the pars reticulata and lateralis. The ventral dendrites descending in the pars reticulata receives inhibitory synapses from the initial axonal collaterals of pars reticulata neurons. Groups of dopaminergic neurons located more dorsally and posteriorly in the tegmentum are of the same type without forming true nuclei. The "cell groups A8 and A10" are spread inside the cerebral peduncule. They are not known to receive striatal afferences and are not in a topographical position to do so. The dopaminergic ensemble is thus also on this point inhomogeneous. This is another major difference with the pallidonigral ensemble. The axons of the dopaminergic neurons, that are thin and varicose, leave the nigra dorsally. They turn round the medial border of the subthalamic nucleus, enter the H2 field above the subthalamic nucleus, then cross the internal capsule to reach the upper part of the medial pallidum where they enter the pallidal laminae, from which they enter the striatum. They end intensively but inhomogeneously in the striatum, rather in the matrix of the anterior part and rather in the striosomes dorsalwards.

Physiology

Contrarily to the neurons of the pars reticulata-lateralis, dopaminergic neurons are "low-spiking pacemakers", spiking at low frequency (0,2 to 10 Hz). The role of the dopaminergic neurons has been the source of a considerable literature. As the pathological disappearance of the black neurons was linked to the appearance of Parkinson's disease, their activity was thought to be "motor". A major discovery has been that the stimulation of the black neurons had no motor effect. Their activity is in fact linked to reward and prediction of reward. In a recent review, it is demonstrated that phasic responses to reward-related events, notably reward-prediction errors, lead to dopamine release" While it is thought that there could be different behavioral processes including long time regulation. Due to its widespread distribution, the dopaminergic system may regulate the basal ganglia system in many places.

Regulators of the Basal Ganglia Core

Subthalamic Nucleus or Corpus Luysi

As indicated by its name, the subthalamic nucleus is located below the thalamus; dorsally to the substantia nigra and medial to the internal capsule. The subthalamic nucleus is lenticular in form and of homogeneous aspect. It is made up of a particular neuronal species having rather long ellipsoid dendritic arborisations, devoid of spines, mimicking the shape of the whole nucleus. The subthalamic neurons are "fast-spiking pacemakers" spiking at 80 to 90 Hz. There are also about 7.5% of GABA microneurons participating in the local circuitry. The subthalamic nucleus receives its main afference from the lateral pallidum. Another afference comes from the cerebral cortex (glutamatergic), particularly from the motor cortex, which is too much neglected in models. A cortical excitation, via the subthalamic nucleus provokes an early short latency excitation leading to an inhibition in pallidal neurons. Subthalamic axons leave the nucleus dorsally. Except for the connection to the striatum (17.3% in macaques), most of the principal neurons are multitargets and ffed axons to the other elements of the core of the basal ganglia. Some send axons to the substantia nigra medially and the medial and lateral nuclei of the pallidum laterally (3-target 21.3%). Some are 2-target with the lateral pallidum and the substantia nigra (2.7%) or the lateral pallidum and the medial(48%). Fewer are single target for the lateral pallidum. If one adds all those reaching this target, the main afference of the subthalamic nucleus is, in 82.7% of the cases, the lateral pallidum (external segment of the globus pallidus. While striatopallidal and the pallido-subthalamic connections are inhibitory (GABA), the subthalamic nucleus utilises the excitatory neurotransmitter glutamate. Its lesion resulting in hemiballismus is known for long. Deep brain stimulation of the nucleus suppress most of the symptoms of the Parkinson' syndrome, particularly dyskinesia induced by dopamine therapy.

Subthalamo-lateropallidal Pacemaker

As said before, the lateral pallidum has purely intrinsic basal ganglia targets. It is particularly linked to the subthalamic nucleus by two-way connections. Contrary to the two output sources (medial pallidum and nigra reticulata), neither the lateral pallidum nor the subthalmic nucleus send axons to the thalamus. The subthalamic nucleus and lateral pallidum are both fast-firing pacemakers. Together they constitute the "central pacemaker of the basal ganglia" with synchronous bursts. The pallido-subthalamic connection is inhibitory, the subthalamo-pallidal is excitatory. They are coupled regulators or coupled autonomous oscillators, the analysis of which has been insufficiently deepened. The lateral pallidum receives a lot of striatal axons, the subthalamic nucleus not. The subthalamic nucleus receives cortical axons, the pallidum not. The subsystem they make with their inputs and outputs corresponds to a classical systemic feedback circuit but it is evidently more complex.

Central Region of the Thalamus

The centromedian nucleus is in the central region of the thalamus. In upper primates it has three parts instead of two, with their own types of neuron. Output from here goes to the subthalamic nucleus and the putamen. Its input includes fibers from the cortex and globus pallidus.

Pedunculopontine Complex

The pedunculopontine nucleus is a part of the reticular formation in the brainstem and a main component of the reticular activating system, and gives a major input to the basal ganglia. As indicated by its name, it is located at the junction between the pons and the cerebral peduncle, and near the substantia nigra. The axons are either excitatory or inhibitory and mainly target the substantia nigra. Another strong input is to the subthalamic nucleus. Other targets are the GPi and the striatum. The complex receives direct afferences from the cortex and above all abundant direct afferences from the medial pallidum (inhibitory). It sends axons to the pallidal territory of the VL. The activity of the neurons is modified by movement, and precede it. All this led Mena-Segovia to propose that the complex be linked in a way or another to the basal ganglia system. A review on its role in the system and in diseases is given by Pahapill and Lozano. It plays an important role in awakeness and sleep. It has a dual role as a regulator of, and of being regulated by the basal ganglia.

Outputs of the Basal Ganglia System

In the cortico-basal ganglia-thalamo-cortical loop the basal ganglia are interconnected, with little output to external targets. One target is the superior colliculus, from the pars reticulata. The two other major output subsystems are to the thalamus and from there to the cortex. In the thalamus the GPimedial fibers are separated from the nigral as their terminal arborisations do not mix. The thalamus relays the nigral output to the premotor and to the frontal cortices.

Medial Pallidum to Thalamic VL and from there to Cortex

The thalamic fasciculus (H1 field) consists of fibers from the ansa lenticularis and from the lenticular fasciculus (H2 field), coming from different portions of the GPi. These tracts are collectively the pallidothalamic tracts and join before they enter the ventral anterior nucleus of the thalamus.

Pallidal axons have their own territory in the ventral lateral nucleus (VL); separated from the cerebellar and nigral territories. The VL is stained for calbindin and acetylcholinesterase. The axons ascend in the nucleus where they branch profusely. The VL output goes preferentially to the supplementary motor cortex (SMA), to the preSMA and to a lesser extent to the motor cortex. The pallidothalamic axons give branches to the pars media of the central complex which sends axons to the premotor and accessory motor cortex.

SNpr to Thalamic VA and from there to Cortex

The ventral anterior nucleus (VA) output targets the premotor cortex, the anterior cingulate cortex and the oculomotor cortex, without significant connection to the motor cortex.

Brow Ridge

The brow ridge, or supraorbital ridge known as superciliary arch in medicine, refers to a bony ridge located above the eye sockets of all primates. In Homo sapiens sapiens (modern humans) the eyebrows are located on their lower margin.

Structure

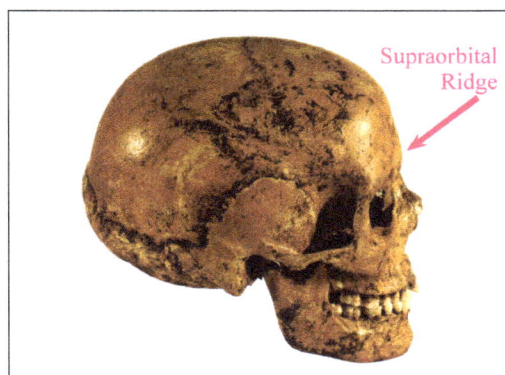

The brow ridges are often not well expressed in human females, as pictured above in a female skull, and are most easily seen in profile.

The brow ridge is a nodule or crest of bone situated on the frontal bone of the skull. It forms the separation between the forehead portion itself (the squama frontalis) and the roof of the eye sockets (the pars orbitalis). Normally, in humans, the ridges arch over each eye, offering mechanical protection. In other primates, the ridge is usually continuous and often straight rather than arched. The ridges are separated from the frontal eminences by a shallow groove. The ridges are most prominent medially, and are joined to one another by a smooth elevation named the glabella.

Typically, the arches are more prominent in men than in women, and vary between different ethnic groups. Behind the ridges, deeper in the bone, are the frontal sinuses.

Terminology

The brow ridges, being a prominent part of the face in some ethnic groups and a trait linked to both atavism and sexual dimorphism, have a number of names in different disciplines. In vernacular English, the terms eyebrow bone or eyebrow ridge are common. The more technical terms frontal or supraorbital arch, ridge or torus (or tori to refer to the plural, as the ridge is usually seen as a pair) are often found in anthropological or archaeological studies. In medicine, the term *arcus superciliaris* (Latin) or the English translation superciliary arch. This feature is different from the supraorbital margin and the margin of the orbit.

Some paleoanthropologists distinguish between frontal torus and supraorbital ridge. In anatomy, a *torus* is a projecting shelf of bone that unlike a ridge is rectilinear, unbroken and goes through glabella. Some fossil hominins, in this use of the word, have the *frontal torus*, but almost all modern humans only have the ridge.

Development

Spatial Model

The Spatial model proposes that supraorbital torus development can be best explained in terms of the disparity between the anterior position of the orbital component relative the neurocranium.

Much of the groundwork for the spatial model was laid down by Schultz. He was the first to document that at later stages of development (after age 4) the growth of the orbit would outpace that of the eye. Consequently, he proposed that facial size is the most influential factor in orbital development, with orbital growth being only secondarily affected by size and ocular position.

Weindenreich and Biegert argued that the supraorbital region can best be understood as a product of the orientation of its two components, the face and the neurocranium.

The most composed articulation of the spatial model was presented by Moss and Young), who stated that "the presence of supraorbital ridges is only the reflection of the spatial relationship between two functionally unrelated cephalic components, the orbit and the brain". They proposed that during infancy the neurocranium extensively overlaps the orbit, a condition that prohibits brow ridge development. As the splanchocranium grows, however, the orbits begin to advance, thus causing the anterior displacement of the face relative to the brain. Brow ridges then form as a result of this separation.

Bio-mechanical Model

The bio-mechanical model predicts that morphological variation in torus size is the direct product of differential tension caused by mastication, as indicated by an increase in load/lever ratio and broad craniofacial angle.

By applying pressure similar to the type associated with chewing, he carried out an analysis of the structural function of the supraorbital region on dry human and gorilla skulls. His findings indicated that the face acts as a pillar that carries and disperses tension caused by the forces produced during mastication. Russell and Oyen elaborated on this idea, suggesting that amplified facial projection necessitates the application of enhanced force to the anterior dentition in order to generate the same bite power that individuals with a dorsal deflection of the facial skull exert. In more prognathic individuals, this increased pressure triggers bone deposition to reinforce the brow ridges, until equilibrium is reached.

Oyen conducted a cross-section study of *Papio anubis* in order to ascertain the relationship between palate length, incisor load and Masseter lever efficiency, relative to torus enlargement. Indications found of osteoblastic deposition in the glabella were used as evidence for supraorbital enlargement. Oyen 's data suggested that more prognathic individuals experienced a decrease in load/lever efficiency. This transmits tension via the frontal process of the maxilla to the supraorbital region, resulting in a contemporary reinforcement of this structure. This was also correlated to periods of tooth eruption.

Russell developed aspects of this mode further. Employing an adult Australian sample, she tested the association between brow ridge formation and anterior dental loading, via the craniofacial angle (prosthion-nasion-metopion), maxilla breadth, and discontinuities in food preparation such as those observed between different age groups. Finding strong support for the first two criteria, she concluded that the supraorbital complex is formed as a result of increased tension due to the widening of the maxilla, thought to be positively correlated with the size of the masseter muscle, as well as with the improper orientation of bone in the superior orbital region.

Function

The brow ridge functions to reinforce the weaker bones of the face in much the same way that the chin of modern humans reinforces their comparatively thin mandibles. This was necessary in pongids and early hominids because of the tremendous strain put on the cranium by their powerful chewing apparatuses, which is best demonstrated by any of the members of the genus *Paranthropus*. The brow ridge was one of the last traits to be lost in the path to anatomically modern humans, and only disappeared in a majority of modern humans with the development of the modern pronounced frontal lobe. This is one of the most salient differences between *Homo sapiens* and other species like the *Homo neanderthalensis*.

Grooming Claw

A grooming claw (or toilet claw) is the specialized claw or nail on the foot of certain primates, used for personal grooming. All prosimians have a grooming claw, but the digit that is specialized in this manner varies. With one possible exception, in the suborder Strepsirrhini, which includes lemurs, galagos and lorises, the grooming claw is on the second toe. The possible exception is the aye-aye, which has claws instead of nails on toes 2 through 5. There is some debate concerning whether any of these claws (and if so which ones) are grooming claws. Less commonly known, a grooming claw is also found on the second pedal digit of night monkeys (Aotus), titis (Callicebus), and possibly other New World monkeys.

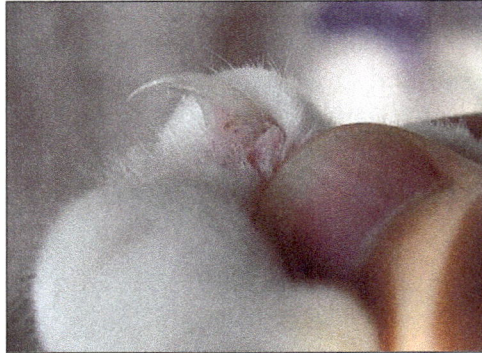

A claw is a curved, pointed appendage, found at the end of a toe or finger in most amniotes.

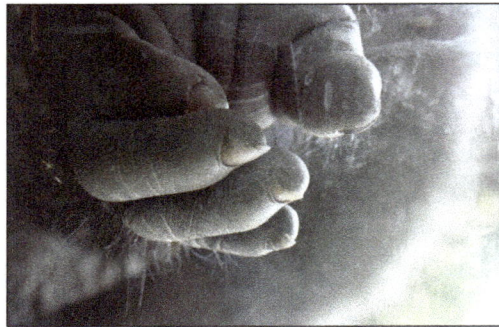

A nail is a horn-like envelope covering the tips of the fingers and toes in most primates and a few other mammals. Nails are similar to claws in other animals. Fingernails and toenails are made of a tough protective protein called alpha-keratin. This protein is also found in the hooves and horns of different animals.

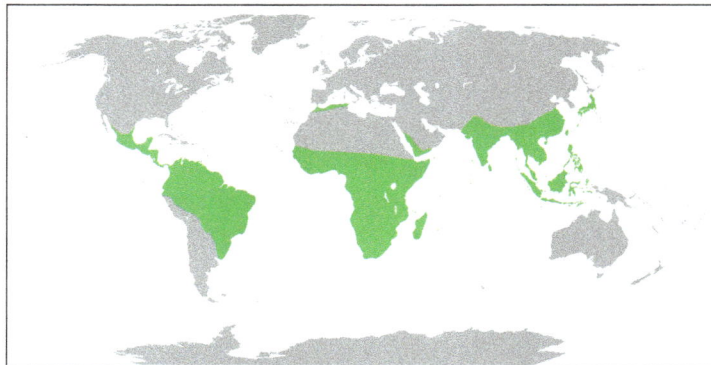

A primate is a eutherian mammal constituting the taxonomic order Primates. Primates arose 85–55 million years ago from small terrestrial mammals (Primatomorpha), which adapted to living in the trees of tropical forests: many primate characteristics represent adaptations to life in this challenging environment, including large brains, visual acuity, color vision, altered shoulder girdle, and dexterous hands. Primates range in size from Madame Berthe's mouse lemur, which weighs 30 g (1 oz), to the eastern gorilla, weighing over 200 kg (440 lb). There are 190–448 species of living primates, depending on which classification is used. New primate species continue to be discovered: over 25 species were described in the first decade of the 2000s, and eleven since 2010.

The first toe is the large one, the equivalent of a human big toe. However, in all these prosimians the foot is more or less hand-like. The first toe is opposable, like a human thumb, and the second and third toes correspond approximately to the index and middle fingers.

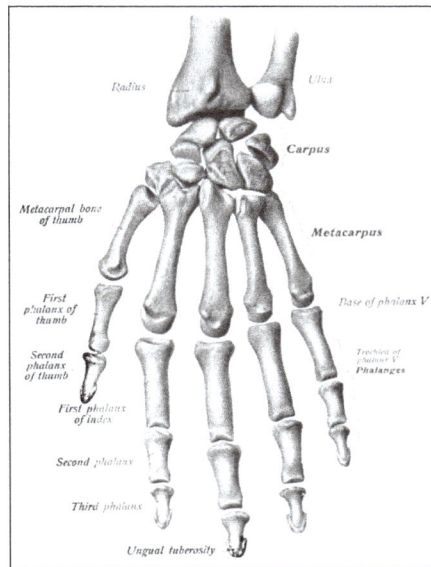

The thumb is the first digit of the hand. When a person is standing in the medical anatomical position, the thumb is the outermost digit. The Medical Latin English noun for thumb is pollex, and the corresponding adjective for thumb is pollical.

Like a claw or a nail, the grooming claw is also made of keratin. It resembles a claw in both its lateral compression and longitudinal curvature. However, the tip is not as pointed, and it always stands at a steeper angle, a characteristic that also distinguishes it from a nail.

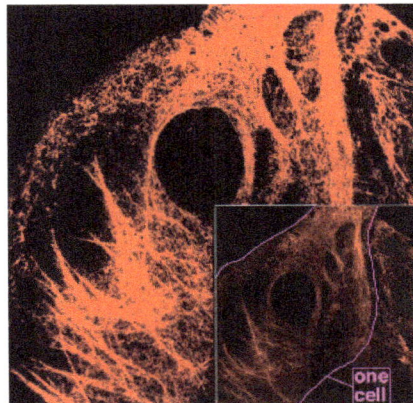

Keratin is one of a family of fibrous structural proteins. It is the key structural material making up hair, nails, horns, claws, hooves, and the outer layer of human skin. Keratin is also the protein that protects epithelial cells from damage or stress. Keratin is extremely insoluble in water and organic solvents. Keratin monomers assemble into bundles to form intermediate filaments, which are tough and form strong unmineralized epidermal appendages found in reptiles, birds, amphibians, and mammals. The only other biological matter known to approximate the toughness of keratinized tissue is chitin.

Function

The grooming claw is used in personal grooming to rake through the fur or scratch, particularly around the head and neck.

Cetacea Anatomy and Physiology

Cetacean bodies are generally similar to that of fish, which can be attributed to their lifestyle and the habitat conditions. Their body is well-adapted to their habitat, although they share essential characteristics with other higher mammals (Eutheria).

They have a streamlined shape, and their forelimbs are flippers. Almost all have a dorsal fin on their backs that can take on many forms depending on the species. A few species, such as the beluga whale, lack them. Both the flipper and the fin are for stabilization and steering in the water.

The male genitals and mammary glands of females are sunken into the body.

The body is wrapped in a thick layer of fat, known as blubber, used for thermal insulation and gives cetaceans their smooth, streamlined body shape. In larger species, it can reach a thickness up to half a meter (1.6 ft).

Sexual dimorphism evolved in many toothed whales. Sperm whales, narwhals, many members of the beaked whale family, several species of the porpoise family, killer whales, pilot whales, eastern spinner dolphins and northern right whale dolphins show this characteristic. Males in these species developed external features absent in females that are advantageous in combat or display. For example, male sperm whales are up to 63% percent larger than females, and many beaked whales possess tusks used in competition among males. Hind legs are not present in cetaceans, nor are any other external body attachments such as a pinna and hair.

Head

Whales have an elongated head, especially baleen whales, due to the wide overhanging jaw. Bowhead whale plates can be 9 metres (30 ft) long. Their nostril(s) make up the blowhole, with one in toothed whales and two in baleen whales.

The nostrils are located on top of the head above the eyes so that the rest of the body can remain submerged while surfacing for air. The back of the skull is significantly shortened and deformed. By shifting the nostrils to the top of the head, the nasal passages extend perpendicularly through the skull. The teeth or baleen in the upper jaw sit exclusively on the maxilla. The braincase is concentrated through the nasal passage to the front and is correspondingly higher, with individual cranial bones that overlap.

In toothed whales, connective tissue exists in the melon as a head buckle. This is filled with air sacs and fat that aid in buoyancy and biosonar. The sperm whale has a particularly pronounced melon; this is called the spermaceti organ and contains the eponymous spermaceti, hence the name "sperm whale". Even the long tusk of the narwhal is a vice-formed tooth. In many toothed whales, the depression in their skull is due to the formation of a large melon and multiple, asymmetric air bags.

River dolphins, unlike most other cetaceans, can turn their head 90°. Other cetaceans have fused neck vertebrae and are unable to turn their head at all.

The baleen of baleen whales consists of long, fibrous strands of keratin. Located in place of the teeth, it has the appearance of a huge fringe and is used to sieve the water for plankton and krill.

Brain

The neocortex of many cetaceans is home to elongated spindle neurons that, prior to 2019, were known only in hominids. In humans, these cells are thought to be involved in social conduct, emotions, judgment and theory of mind. Cetacean spindle neurons are found in areas of the brain homologous to where they are found in humans, suggesting they perform a similar function.

Brain size was previously considered a major indicator of intelligence. Since most of the brain is used for maintaining bodily functions, greater ratios of brain to body mass may increase the amount of brain mass available for cognitive tasks. Allometric analysis indicates that mammalian brain size scales at approximately two-thirds or three-quarter exponent of the body mass. Comparison of a particular animal's brain size with the expected brain size based on such an analysis provides an encephalization quotient that can be used as an indication of animal intelligence. Sperm whales have the largest brain mass of any animal on earth, averaging 8,000 cm^3 (490 in^3) and 7.8 kg (17 lb) in mature males. The brain to body mass ratio in some odontocetes, such as belugas and narwhals, is second only to humans. In some whales, however, it is less than half that of humans: 0.9% versus 2.1%. The sperm whale (*Physeter macrocephalus*) is the largest of all toothed predatory animals and possesses the largest brain.

Skeleton

Cetacea Skeletons

Skeleton of a blue whale standing outside the Long Marine.

Weathered upper jaw of a sperm whale.

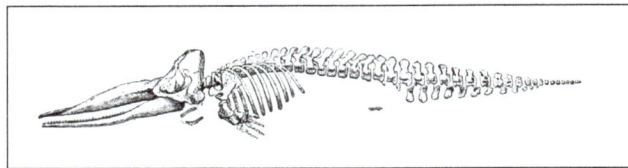
Sperm whale skeleton.

The cetacean skeleton is largely made up of cortical bone, which stabilizes the animal in the water. For this reason, the usual terrestrial compact bones, which are finely woven cancellous bone, are replaced with lighter and more elastic material. In many places, bone elements are replaced by cartilage and even fat, thereby improving their hydrostatic qualities. The ear and the muzzle contain a bone shape that is exclusive to cetaceans with a high density, resembling porcelain. This conducts sound better than other bones, thus aiding biosonar.

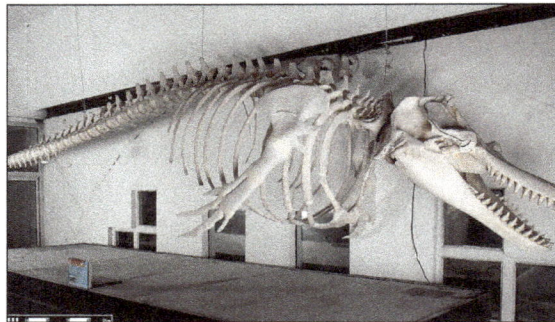
Killer whale skeleton.

The number of vertebrae that make up the spine varies by species, ranging from forty to ninety-three. The cervical spine, found in all mammals, consists of seven vertebrae which, however, are reduced or fused. This fusion provides stability during swimming at the expense of mobility. The fins are carried by the thoracic vertebrae, ranging from nine to seventeen individual vertebrae. The sternum is cartilaginous. The last two to three pairs of ribs are not connected and hang freely in the body wall. The stable lumbar and tail include the other vertebrae. Below the caudal vertebrae is the chevron bone.

The front limbs are paddle-shaped with shortened arms and elongated finger bones, to support movement. They are connected by cartilage. The second and third fingers display a proliferation of the finger members, a so-called hyperphalangy. The shoulder joint is the only functional joint in all cetaceans except for the Amazon river dolphin. The collarbone is completely absent.

Fluke

Humpback whale fluke.

They have a cartilaginous fluke at the end of their tails that is used for propulsion. The fluke is set horizontally on the body, unlike fish, which have vertical tails.

Physiology

Circulation

Cetaceans have powerful hearts. Blood oxygen is distributed effectively throughout the body. They are warm-blooded, i.e., they hold a nearly constant body temperature.

Respiration

Cetaceans have lungs, meaning they breathe air. An individual can last without a breath from a few minutes to over two hours depending on the species. Cetacea are deliberate breathers who must be awake to inhale and exhale. When stale air, warmed from the lungs, is exhaled, it condenses as it meets colder external air. As with a terrestrial mammal breathing out on a cold day, a small cloud of 'steam' appears. This is called the 'spout' and varies across species in shape, angle and height. Species can be identified at a distance using this characteristic.

The structure of the respiratory and circulatory systems is of particular importance for the life of marine mammals. The oxygen balance is effective. Each breath can replace up to 90% of the total lung volume. For land mammals, in comparison, this value is usually about 15%. During inhalation, about twice as much oxygen is absorbed by the lung tissue as in a land mammal. As with all mammals, the oxygen is stored in the blood and the lungs, but in cetaceans, it is also stored in various tissues, mainly in the muscles. The muscle pigment, myoglobin, provides an effective bond. This additional oxygen storage is vital for deep diving, since beyond a depth around 100 m (330 ft), the lung tissue is almost completely compressed by the water pressure.

Organs

The stomach consists of three chambers. The first region is formed by a loose gland and a muscular forestomach (missing in beaked whales), which is then followed by the main stomach and the pylorus. Both are equipped with glands to help digestion. A bowel adjoins the stomachs, whose individual sections can only be distinguished histologically. The liver is large and separate from the gall bladder.

The kidneys are long and flattened. The salt concentration in cetacean blood is lower than that in seawater, requiring kidneys to excrete salt. This allows the animals to drink seawater.

Senses

Cetacean eyes are set on the sides rather than the front of the head. This means only species with pointed 'beaks' (such as dolphins) have good binocular vision forward and downward. Tear glands secrete greasy tears, which protect the eyes from the salt in the water. The lens is almost spherical, which is most efficient at focusing the minimal light that reaches deep water. Cetaceans are known to possess excellent hearing.

At least one species, the tucuxi or Guiana dolphin, is able to use electroreception to sense prey.

Ears

Biosonar.

The external ear has lost the pinna (visible ear), but still retains a narrow external auditory meatus. To register sounds, instead, the posterior part of the mandible has a thin lateral wall (the pan bone) fronting a concavity that houses a fat pad. The pad passes anteriorly into the greatly enlarged mandibular foramen to reach in under the teeth and posteriorly to reach the thin lateral wall of the ectotympanic. The ectotympanic offers a reduced attachment area for the tympanic membrane. The connection between this auditory complex and the rest of the skull is reduced to a single, small cartilage in oceanic dolphins.

In odontocetes, the complex is surrounded by spongy tissue filled with air spaces, while in mysticetes, it is integrated into the skull as with land mammals. In odontocetes, the tympanic membrane (or ligament) has the shape of a folded-in umbrella that stretches from the ectotympanic ring and narrows off to the malleus (quite unlike the flat, circular membrane found in land mammals.) In mysticetes, it also forms a large protrusion (known as the "glove finger"), which stretches into the external meatus and the stapes are larger than in odontocetes. In some small sperm whales, the malleus is fused with the ectotympanic.

The ear ossicles are pachyosteosclerotic (dense and compact) and differently shaped from land mammals (other aquatic mammals, such as sirenians and earless seals, have also lost their pinnae). T semicircular canals are much smaller relative to body size than in other mammals.

The auditory bulla is separated from the skull and composed of two compact and dense bones (the periotic and tympanic) referred to as the tympanoperiotic complex. This complex is located in a cavity in the middle ear, which, in the Mysticeti, is divided by a bony projection and compressed between the exoccipital and squamosal, but in the odontoceti, is large and completely surrounds the bulla (hence called "peribullar"), which is, therefore, not connected to the skull except in physeterids. In the Odontoceti, the cavity is filled with a dense foam in which the bulla hangs suspended in five or more sets of ligaments. The pterygoid and peribullar sinuses that form the cavity tend to be more developed in shallow water and riverine species than in pelagic Mysticeti. In

Odontoceti, the composite auditory structure is thought to serve as an acoustic isolator, analogous to the lamellar construction found in the temporal bone in bats.

Cetaceans use sound to communicate, using groans, moans, whistles, clicks or the 'singing' of the humpback whale.

Echolocation

Odontoceti are generally capable of echolocation. They can discern the size, shape, surface characteristics, distance and movement of an object. They can search for, chase and catch fast-swimming prey in total darkness. Most Odontoceti can distinguish between prey and nonprey (such as humans or boats); captive Odontoceti can be trained to distinguish between, for example, balls of different sizes or shapes. Echolocation clicks also contain characteristic details unique to each animal, which may suggest that toothed whales can discern between their own click and that of others.

Mysticeti have exceptionally thin, wide basilar membranes in their cochleae without stiffening agents, making their ears adapted for processing low to infrasonic frequencies.

Chromosomes

The initial karyotype includes a set of chromosomes from $2n = 44$. They have four pairs of telocentric chromosomes (whose centromeres sit at one of the telomeres), two to four pairs of subtelocentric and one or two large pairs of submetacentric chromosomes. The remaining chromosomes are metacentric the centromere is approximately in the middle and are rather small. Sperm whales, beaked whales and right whales converge to a reduction in the number of chromosomes to $2n = 42$.

Equine Anatomy and Physiology

Equine anatomy refers to the gross and microscopic anatomy of horses and other equids, including donkeys, and zebras. While all anatomical features of equids are described in the same terms as for other animals by the International Committee on Veterinary Gross Anatomical Nomenclature, there are many horse-specific colloquial terms used by equestrians.

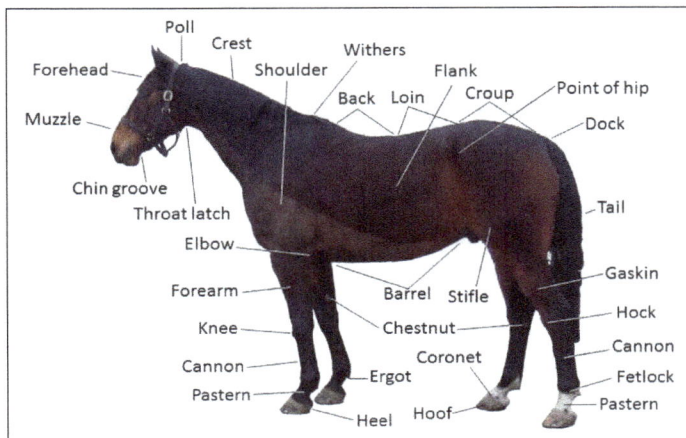

Points of a horse.

External Anatomy

- Back: The area where the saddle sits, beginning at the end of the withers, extending to the last thoracic vertebrae (colloquially includes the loin or "coupling," though technically incorrect usage).

- Barrel: The body of the horse, enclosing the rib cage and the major internal organs.

- Buttock: The part of the hindquarters behind the thighs and below the root of the tail.

- Cannon or cannon bone: The area between the knee or hock and the fetlock joint, sometimes called the "shin" of the horse, though technically it is the metacarpal III.

- Chestnut: A callosity on the inside of each leg.

- Chin groove: The part of the horse's head behind the lower lip and chin, the area that dips down slightly on the lower jaw; area where the curb chain of certain bits is fastened.

- Coronet or coronary band: The ring of soft tissue just above the horny hoof that blends into the skin of the leg.

- Crest: The upper portion of the neck where the mane grows.

- Croup: The topline of the hindquarters, beginning at the hip, extending proximate to the sacral vertebrae and stopping at the dock of the tail (where the coccygeal vertebrae begin); sometimes called "rump".

- Dock: The living part of the tail, consisting of the coccygeal vertebrae, muscles and ligaments. Sometimes used colloquially to refer to the root of the tail, below.

- Elbow: The joint of the front leg at the point where the belly of the horse meets the leg. Homologous to the elbow in humans.

- Ergot: A callosity on the back of the fetlock.

- Face: The area between the forehead and the tip of the upper lip.

- Fetlock: Sometimes called the "ankle" of the horse, though it is not the same skeletal structure as an ankle in humans; known to anatomists as the metacarpophalangeal (front) or metatarsophalangeal (hind) joint; homologous to the "ball" of the foot or the metacarpophalangeal joints of the fingers in humans.

- Flank: Where the hind legs and the barrel meet, specifically the area right behind the rib cage and in front of the stifle joint.

- Forearm: The area of the front leg between the knee and elbow, consisting of the fused radius and ulna, and all the tissue around these bones; anatomically, the antebrachium.

- Forehead: The area between the poll, the eyes and the arch of the nose.

- Forelock: The continuation of the mane, which hangs from between the ears down onto the forehead of the horse.

- Frog: The highly elastic wedge-shaped mass on the underside of the hoof, which normally makes contact with the ground every stride, and supports both the locomotion and circulation of the horse.

- Gaskin: The large muscle on the hind leg, just above the hock, below the stifle, homologous to the calf of a human.

- Girth or heartgirth: The area right behind the elbow of the horse, where the girth of the saddle would go; this area should be where the barrel is at its greatest diameter in a properly-conditioned horse that is not pregnant or obese.

- Hindquarters: The large, muscular area of the hind legs, above the stifle and behind the barrel.

- Hock: The tarsus of the horse (hindlimb equivalent to the human ankle and heel), the large joint on the hind leg.

- Hoof: The foot of the horse; the hoof wall is the tough outside covering of the hoof that comes into contact with the ground and is, in many respects, a much larger and stronger version of the human fingernail.

- Jugular Groove: The line of indentation on the lower portion of the neck, can be seen from either side, just above the windpipe; beneath this area run the jugular vein, the carotid artery and part of the sympathetic trunk.

- Knee: The carpus of the horse (equivalent to the human wrist), the large joint in the front legs, above the cannon bone.

- Loin: The area right behind the saddle, going from the last rib to the croup, anatomically approximate to the lumbar spine.

- Mane: Long and relatively coarse hair growing from the dorsal ridge of the neck.

- Muzzle: The chin, mouth, and nostrils of the face.

- Pastern: The connection between the coronet and the fetlock, made up of the middle and proximal phalanx.

- Poll: Commonly refers to the poll joint at the beginning of the neck, immediately behind the ears, a slight depression at the joint where the atlas (C1) meets the occipital crest; anatomically, the occipital crest itself is the "poll".

- Root of the tail or root of the dock: The point where the tail is "set on" (attached) to the rump; Sometimes also called the "dock".

- Shoulder: Made up of the scapula and associated muscles, runs from the withers to the point of shoulder (the joint at the front of the chest, i.e. the glenoid); the angle of the shoulder has a great effect on the horse's movement and jumping ability, and is an important aspect of equine conformation.

- Splints: Bones found on each of the legs, on either side of the cannon bone (8 total); partially vestigial, these bones support the corresponding carpal bones in the forelimb, and the corresponding tarsal bones in the hindlimb; anatomically referred to as Metacarpal/Metatarsal II (on the medial aspect (inside)) and IV (on the lateral aspect (outside)).

- Stifle: Corresponds to the knee of a human, consists of the articulation between femur and tibia, as well as the articulation between patella and femur.

- Tail: The long hairs which grow from the dock; may also include the dock.

- Throatlatch (also, throttle, throatlash, throat): The point at which the windpipe meets the head at the underside of the jaw, corresponding to where the eponymous part of a bridle goes.

- Withers: The highest point of the thoracic vertebrae, the point just above the tops of the shoulder blades, seen best with horse standing square and head slightly lowered; the height of the horse is measured at the withers.

Digestive System

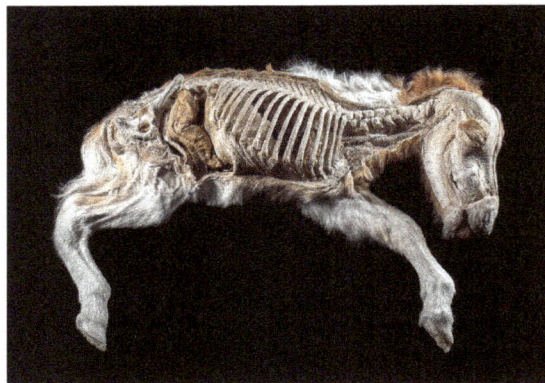

A dehydrated anatomical specimen.

Horses and other equids evolved as grazing animals, adapted to eating small amounts of the same kind of food all day long. In the wild, the horse adapted to eating prairie grasses in semi-arid regions and traveling significant distances each day in order to obtain adequate nutrition. Therefore, the digestive system of a horse is about 30 m (100 ft) long, and most of this is intestines.

Mouth

Digestion begins in the mouth, which is also called the "oral cavity." It is made up of the teeth, the hard palate, the soft palate, the tongue and related muscles, the cheeks and the lips. Horses also have three pairs of salivary glands, the parotoid (largest salivary gland and located near the poll), mandibular (located in the jaw), and sublingual (located under the tongue). Horses select pieces of forage and pick up finer foods, such as grain, with their sensitive, prehensile lips. The front teeth of the horse, called incisors, clip forage, and food is then pushed back in the mouth by the tongue, and ground up for swallowing by the premolars and molars.

Esophagus

The esophagus is about 1.2 to 1.5 m (4 to 5 ft) in length, and carries food to the stomach. A muscular ring, called the cardiac sphincter, connects the stomach to the esophagus. This sphincter is very well developed in horses. This and the oblique angle at which the esophagus connects to the stomach explains why horses cannot vomit. The esophagus is also the area of the digestive tract where horses may suffer from choke.

Stomach

Equine stomach.

Horses have a relatively small stomach for their size, and this limits the amount of feed a horse can take in at one time. The average sized horse (360 to 540 kg [800 to 1,200 lb]) has a stomach with a capacity of around 19 L (5 US gal), and works best when it contains about 7.6 L (2 US gal). Because the stomach empties when $\frac{2}{3}$ full, whether stomach enzymes have completed their processing of the food or not, and doing so prevents full digestion and proper utilization of feed, continuous foraging or several small feedings per day are preferable to one or two large ones. The horse stomach consists of a non-glandular proximal region (saccus cecus), divided by a distinct border, the margo plicatus, from the glandular distal stomach.

In the stomach, assorted acids and the enzyme pepsin break down food. Pepsin allows for the further breakdown of proteins into amino acid chains. Other enzymes include resin and lipase. Additionally, the stomach absorbs some water, as well as ions and lipid-soluble compounds.

Small Intestine

The horse's small intestine is 15 to 21 m (50 to 70 ft) long and holds 38 to 45 L (10 to 12 US gal). This is the major digestive organ, and where most nutrients are absorbed. It has three parts, the duodenum, jejunum and ileum. The majority of digestion occurs in the duodenum while the majority of absorption occurs in the jejunum. Bile from the liver aids in digesting fats in the duodenum combined with enzymes from the pancreas and small intestine. Horses do not have a gall bladder, so bile flows constantly. Most food is digested and absorbed into the bloodstream from the small intestine, including proteins, simple carbohydrate, fats, and vitamins A, D, and E. Any remaining liquids and roughage move into the large intestine.

Large Intestine

Equine colon.

Cecum

The cecum is the first section of the large intestine. It is also known as the "water gut" or "hind gut". It is a cul-de-sac pouch, about 1.2 m (4 ft) long that holds 26 to 30 L (7 to 8 US gal). It contains bacteria that digest cellulose plant fiber through fermentation. These bacteria feed upon chyme digestive, and also produce certain fat-soluble vitamins which are absorbed by the horse. The reason horses must have their diets changed slowly is so the bacteria in the cecum are able to modify and adapt to the different chemical structure of new feedstuffs. Too abrupt a change in diet can cause colic, as the new food is not properly digested.

Other Section of the Large Intestine

The large colon, small colon, and rectum make up the remainder of the large intestine. The large colon is 3.0 to 3.7 m (10 to 12 ft) long and holds up to 76 L (20 US gal) of semi-liquid matter. It is made up of the right ventral (lower) colon, the left ventral colon, the left dorsal (upper) colon, the right dorsal colon, and the transverse colon, in that order. Three flexures are also named; the sternal flexure, between right and left ventral colon; the pelvic flexure, between left ventral and left dorsal colon; the diaphragmatic flexure, between left dorsal and right dorsal colon. The main purpose of the large colon is to absorb carbohydrates, which were broken down from cellulose in the cecum. Due to its many twists and turns, it is a common place for a type of horse colic called an impaction.

The small colon is 3.0 to 3.7 m (10 to 12 ft) in length and holds only 19 L (5 US gal) of material. It is the area where the majority of water in the horse's diet is absorbed, and is the place where fecal lumps are formed. The rectum is about 30 cm (1 ft) long, and acts as a holding chamber for waste matter, which is then expelled from the body via the anus.

Reproductive System

Mare

The mare's reproductive system is responsible for controlling gestation, birth, and lactation, as well as her estrous cycle and mating behavior. It lies ventral to the 4th or 5th lumbar vertebrae, although its position within the mare can vary depending on the movement of the intestines and distention of the bladder.

The mare has two ovaries, usually 7 to 8 cm (2.8 to 3.1 in) in length and 3 to 4 cm (1.2 to 1.6 in) thick, that generally tend to decrease in size as the mare ages. In equine ovaries, unlike in humans, the vascular tissue is cortical to follicular tissue, so ovulation can only occur at an ovulation fossa near the infundibulum. The ovaries connect to the fallopian tubes (oviducts), which serve to move the ovum from the ovary to the uterus. To do so, the oviducts are lined with a layer of cilia, which produce a current that flows toward the uterus. Each oviduct attaches to one of the two horns of the uterus, which are approximately 20 to 25 cm (7.9 to 9.8 in) in length. These horns attach to the body of the uterus (18 to 20 cm [7.1 to 7.9 in] long). The equine uterus is bipartite, meaning the two uterine horns fuse into a relatively large uterine body (resembling a shortened bicornuate uterus or a stretched simplex uterus). Caudal to the uterus is the cervix, about 5 to 7 cm (2.0 to 2.8 in) long, which separates the uterus from the vagina. Usually 3.5 to 4 cm (1.4 to 1.6 in) in diameter with longitudinal folds on the interior surface, it can expand to allow the passage of the foal. The vagina of the mare is 15 to 20 cm (5.9 to 7.9 in) long, and is quite elastic, allowing it to expand. The vulva is the external opening of the vagina, and consists of the clitoris and two labia. It lies ventral to the rectum. The mare has two mammary glands, which are smaller in maiden mares. They have two ducts each, which open externally.

Stallion

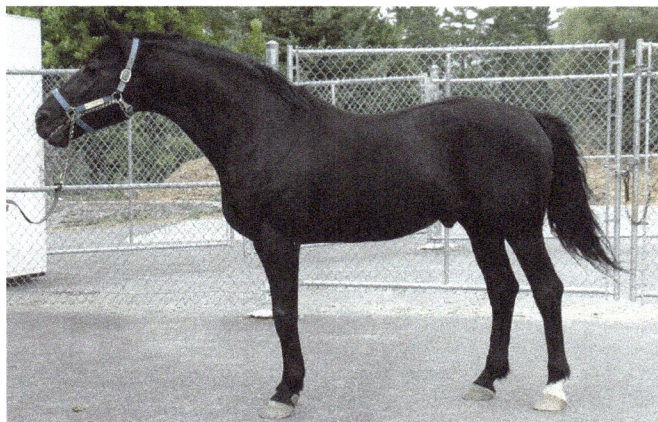

Secondary characteristics of a stallion include heavier muscling for a given breed than is seen in mares or geldings, often with considerable development along the crest of the neck.

The stallion's reproductive system is responsible for his sexual behavior and secondary sex characteristics (such as a large crest). The external genitalia include the urethra; the testes, which average 8 to 12 cm (3.1 to 4.7 in) long; the penis, which, when housed within the prepuce, is 50 cm (20 in) long and 2.5 to 6 cm (0.98 to 2.36 in) in diameter with the distal end 15 to 20 cm (5.9 to 7.9 in) and when erect, increases by 3 to 4 times. The internal genitalia accessory sex glands are the vesicular glands, prostate gland, and bulbourethral glands, which contribute fluid to the semen at ejaculation, but are not strictly necessary for fertility.

Teeth

A horse's teeth include incisors, premolars, molars, and sometimes canine teeth. A horse's incisors, premolars, and molars, once fully developed, continue to erupt throughout its lifetime as the grinding surface is worn down through chewing. Because of this pattern of wear, a rough estimate of a horse's age can be made from an examination of the teeth. Abnormal wear of the teeth, caused by conformational defects, abnormal behaviors, or improper diets, can cause serious health issues and can even result in the death of the horse.

Feet/Hooves

The hoof of the horse encases the second and third phalanx of the lower limbs, analogous to the fingertip or toe tip of a human. In essence, a horse travels on its "tiptoes". The hoof wall is a much larger, thicker and stronger version of the human fingernail or toenail, made up of similar materials, primarily keratin, a very strong protein molecule. The horse's hoof contains a high proportion of sulfur-containing amino acids which contribute to its resilience and toughness. Vascular fold-like structures called laminae suspend the distal phalanx from the hoof wall.

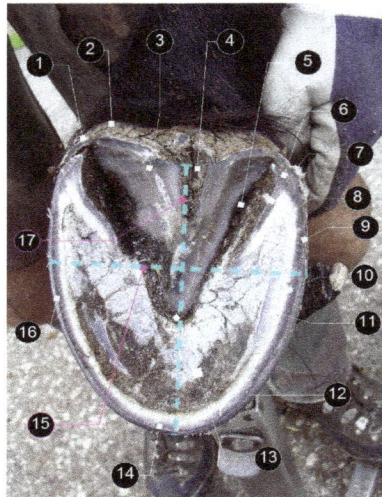

1- Heel perioplium, 2-Bulb, 3-Frog, 4-Frog cleft, 5-Lateral groove, 6-Heel, 7-Bar, 8-Seat-of-corn, 9-Pigmented walls 10-Water line, 11-White line, 12-Apex of the frog, 13-Sole, 14-Toe, 15-How to measure hoof width (blue dotted line), 16-Quarter, 17-How to measure length (blue dotted line).

Skeletal System

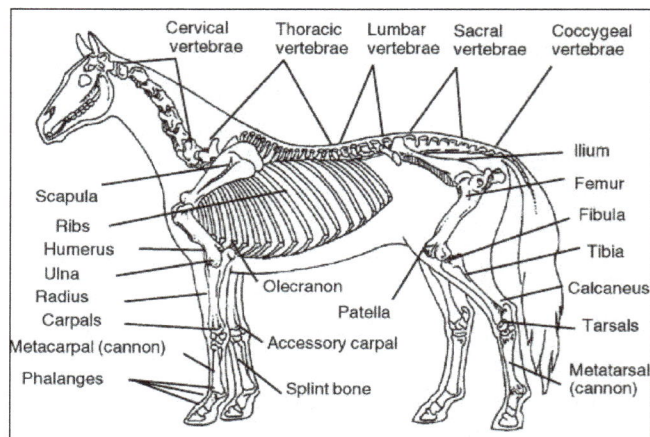

A horse's skeleton.

The skeleton of the horse has three major functions in the body. It protects vital organs, provides framework, and supports soft parts of the body. Horses have 205 bones, which are divided into the appendicular skeleton (the legs) and the axial skeleton (the skull, vertebral column, sternum, and ribs). Both pelvic and thoracic limbs contain the same number of bones, 20 bones per limb. Bones are connected to muscles via tendons and other bones via ligaments. Bones are also used to store minerals, and are the site of red blood cell formation.

- The Appendicular system includes the limbs of the horse.

- The Axial system is composed of the spine, ribs and skull.

The bones of the horse are the same as those of other domestic species, but the third metacarpal and metatarsal are much more developed and the second and fourth are undeveloped, having the first and fifth metacarpal and metatarsal.

Horse Skeleton Bones

Spine	54
Ribs	36
Sternum	01
Head (including ear)	34
Thoracic region	40
Pelvic region	40

Ligaments and Tendons

Ligaments

Ligaments attach bone to bone or bone to tendon, and are vital in stabilizing joints as well as sup-porting structures. They are made up of fibrous material that is generally quite strong. Due to their relatively poor blood supply, ligament injuries generally take a long time to heal.

Tendons

Tendons are cords of connective tissue attaching muscle to bone, cartilage or other tendons. They are a major contributor to shock absorption, are necessary for support of the horse's body, and translate the force generated by muscles into movement. Tendons are classified as flexors (flex a joint) or extensors (extend a joint). However, some tendons will flex multiple joints while extend-ing another (the flexor tendons of the hind limb, for example, will flex the fetlock, pastern, and cof-fin joint, but extend the hock joint). In this case, the tendons (and associated muscles) are named for their most distal action (digital flexion).

Tendons form in the embryo from fibroblasts which become more tightly packed as the tendon grows. As tendons develop they lay down collagen, which is the main structural protein of connec-tive tissue. As tendons pass near bony prominences, they are protected by a fluid filled synovial structure, either a tendon sheath or a sac called a bursa.

Tendons are easily damaged if placed under too much strain, which can result in a painful, and possibly career-ending, injury. Tendinitis is most commonly seen in high performance horses that gallop or jump. When a tendon is damaged the healing process is slow because tendons have a poor blood supply, reducing the availability of nutrients and oxygen to the tendon. Once a tendon is damaged the tendon will always be weaker, because the collagen fibres tend to line up in ran-dom arrangements instead of the stronger linear pattern. Scar tissue within the tendon decreases the overall elasticity in the damaged section of the tendon as well, causing an increase in strain on adjacent uninjured tissue.

Muscular System

When a muscle contracts, it pulls a tendon, which acts on the horse's bones to move them. Muscles are commonly arranged in pairs so that they oppose each other (they are "antagonists"), with one flexing the joint (a flexor muscle) and the other extending it (extensor muscle). Therefore, one muscle of the pair must be relaxed in order for the other muscle in the pair to contract and bend the joint properly. A muscle is made up of several muscle bundles, which in turn are made up of muscle fibers. Muscle fibers have myofibrils, which are able to contract due to actin and myosin. A muscle together with its tendon and bony attachments form an extensor or flexor unit.

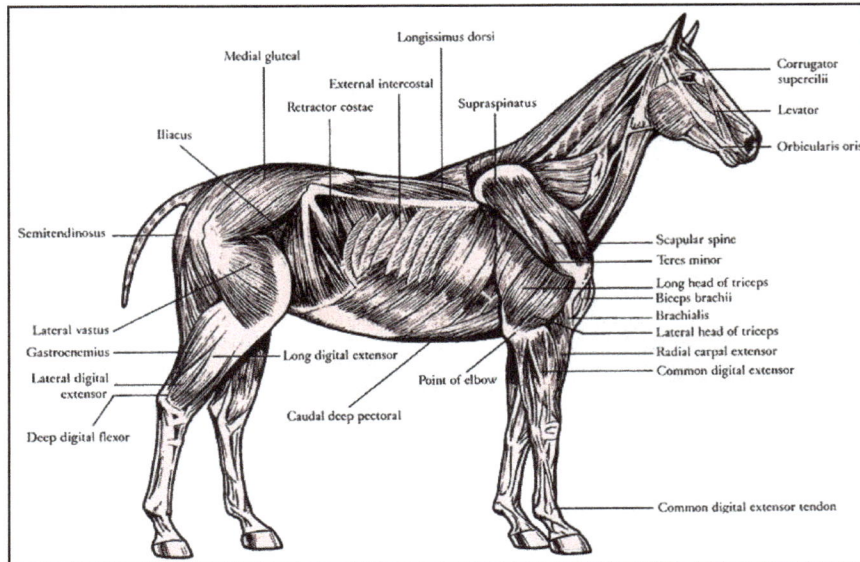

Respiratory System and Smell

The horse's respiratory system consists of the nostrils, pharynx, larynx, trachea, diaphragm, and lungs. Additionally, the nasolacrimal duct and sinuses are connected to the nasal passage. The horse's respiratory system not only allows the animal to breathe, but also is important in the horse's sense of smell (olfactory ability) as well as in communicating. The soft palate blocks off the pharynx from the mouth (oral cavity) of the horse, except when swallowing. This helps prevent the horse from inhaling food, but also means that a horse cannot use its mouth to breathe when in respiratory distress—a horse can only breathe through its nostrils, also called obligate nasal breathing. For this same reason, horses also cannot pant as a method of thermoregulation. The genus *Equus* also has a unique part of the respiratory system called the guttural pouch, which is thought to equalize air pressure on the tympanic membrane. Located between the mandibles but below the occiput, it fills with air when the horse swallows or exhales.

Circulatory System

The horse's circulatory system includes the four-chambered heart, averaging 3.9 kg (8.5 lb) in weight, as well as the blood and blood vessels. Its main purpose is to circulate blood throughout the body to deliver oxygen and nutrients to tissues, and to remove waste from these tissues. The hoof (including the frog - the V shaped part on the bottom of the horses hoof) is a very important part of the circulatory system. As the horse puts weight onto the hoof, the hoof wall is pushed outwards and the frog compressed, driving blood out of the frog, the digital pad, and the laminae of the hoof. When weight is removed from the hoof, the release of pressure pulls blood back down into the foot again. This effectively creates an auxiliary blood-pumping system at the end of each leg. Some of this effect may be lost when a horse is shod (eliminating the expansion and contraction of the hoof wall and raising the frog higher from the ground).

The Eye

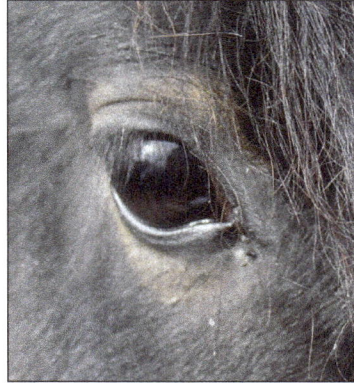

A horse's eye.

The horse has one of the largest eyes of all land mammals. Eye size in mammals is significantly correlated to maximum running speed as well as to body size, in accordance with Leuckart's law; animals capable of fast locomotion require large eyes. The eye of the horse is set to the side of its skull, consistent with that of a prey animal. The horse has a wide field of monocular vision, as well as good visual acuity. Horses have two-color, or dichromatic vision, which is somewhat like red-green color blindness in humans. Because the horse's vision is closely tied to behavior, the horse's visual abilities are often taken into account when handling and training the animal.

Hearing

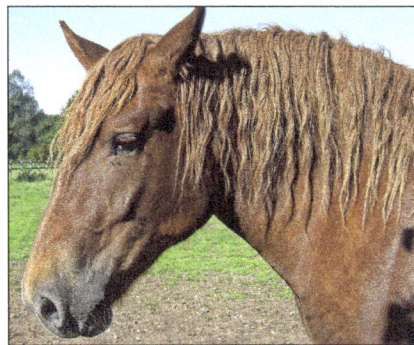

The pinna of a horse's ears can rotate in any direction to pick up sounds.

The hearing of horses is good, superior to that of humans, and the pinna of each ear can rotate up to 180°, giving the potential for 360° hearing without having to move the head. Often, the eye of the horse is looking in the same direction as the ear is directed.

Dog Anatomy and Physiology

Dog anatomy comprises the anatomical studies of the visible parts of the body of a canine. Details of structures vary tremendously from breed to breed, more than in any other animal species, wild or domesticated, as dogs are highly variable in height and weight. The smallest known adult dog was a Yorkshire Terrier that stood only 6.3 cm (2.5 in) at the shoulder, 9.5 cm (3.7 in) in length along the head and body, and weighed only 113 grams (4.0 oz). The largest known adult dog was an English Mastiff which weighed 155.6 kg (343 lb) and was 250 cm (98 in) from the snout to the tail. The tallest known adult dog is a Great Dane that stands 106.7 cm (42.0 in) at the shoulder.

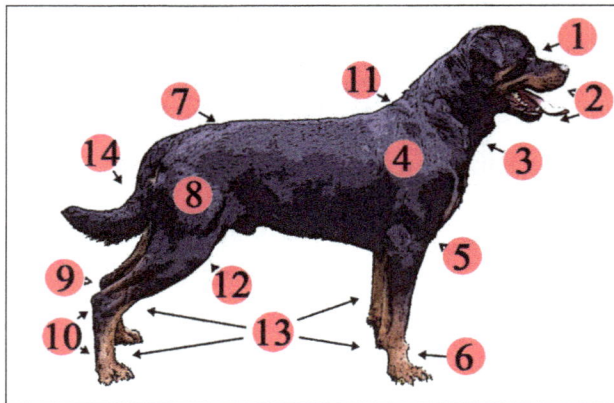

External anatomy (topography) of a typical dog: 1. Stop 2. Muzzle 3. Dewlap (throat, neck skin) 4. Shoulder 5. Elbow 6. Forefeet 7. Croup (rump) 8. Leg (thigh and hip) 9. Hock 10. Hind feet 11. Withers 12. Stifle 13. Paws 14. Tail.

Anatomy

Muscles

The following is a list of the muscles in the dog along with their origin, insertion, action and innervation.

Extrinsic muscles of the thoracic limb and related structures:

Descending superficial pectoral: Originates on the first sternebrae and inserts on the greater tubercle of the humerus. It both adducts the limb and also prevents the limb from being abducted during weight bearing. It is innervated by the cranial pectoral nerves.

Transverse superficial pectoral: Originates on the second and third sternebrae and inserts on the greater tubercle of the humerus. It also adducts the limb and prevents the limb from being abducted during weight bearing. It is innervated by the cranial pectoral nerves.

Deep pectoral: Originates on the ventral sternum and inserts on the lesser tubercle of the humerus. It acts to extend the shoulder joint during weight bearing and flexes the shoulder when there is no weight. It is innervated by the caudal pectoral nerves.

Sternocephalicus: Originates on the sternum and inserts on the temporal bone of the head. Its function is to move the head and neck from side to side. It is innervated by the accessory nerve.

Sternohyoideus: Originates on the sternum and inserts on the basihyoid bone. Its function is to move the tongue caudally. It is innervated by the ventral branches of the cervical spinal nerves.

Sternothyoideus: Originates on the first coastal cartilage and inserts on the thyroid cartilage. Its function is also to move the tongue caudally. It is innervated by the ventral branches of the cervical spinal nerves.

Omotransversarius: Originates on the spine of the scapula and inserts on the wing of the atlas. Its function is to advance the limb and flex the neck laterally. It is innervated by the accessory nerve.

Trapezius: Originates on the supraspinous ligament and inserts on the spine of the scapula. Its function is to elevate and abduct the forelimb. It is innervated by the accessory nerve.

Rhomboideus: Originates on the nuchal crest of the occipital bone and inserts on the scapula. Its function is to elevate the forelimb. It is innervated by the ventral branches of the spinal nerves.

Latissimus dorsi: Originates on thoracolumbar fascia and inserts on the teres major tuberosity of the humerus. Its function is to flex the shoulder joint. It is innervated by the thoracodorsal nerve.

Serratus ventralis: Originates on the transverse processes of the last 5 cervical vertebrae and inserts on the scapula. Its function is to support the trunk and depress the scapula. It is innervated by the ventral branches of the cervical spinal nerves.

Intrinsic muscles of the thoracic limb:

Deltoideus: Originates on the acromial process of the scapula and inserts on the deltoid tuberosity. It acts to flex the shoulder. It is innervated by the axillary nerve.

Infraspinatus: Originates on the infraspinatus fossa and inserts on the greater tubercle of the humerus. It acts to Extend and flex the shoulder joint. It is innervated by the suprascapular nerve.

Teres minor: Originates on the infra glenoid tubercle on the scapula and inserts on the teres minor tuberosity of the humerus. It acts to flex the shoulder and rotate the arm laterally. It is innervated by the axillary nerve.

Supraspinatus: Originates on the supraspinous fossa and inserts on the greater tubercle of the humerus. It acts to extend and stabilize the shoulder joint. It is innervated by the suprascapular nerve.

Medial muscles of the scapula and shoulder:

Subscapularis: Originates on the subscapular fossa and inserts on the greater tubercle of the humerus. It acts to rotate the arm medially and stabilize the joint. It is innervated by the subscapular nerve.

Teres major: Originates on the scapula and inserts on the teres major tuberosity of the humerus. It acts to flex the shoulder and rotate the arm medially. It is innervated by the axillary nerve.

Coracobrachialis: Originates on the coracoid process of the scapula and inserts on the crest of the lesser tubercle of the humerus. It acts to adduct, extend and stabilize the shoulder joint. It is innervated by the musculocutaneous nerve.

Caudal muscles of brachium:

Tensor fasciae antebrachium: Originates on the fascia covering the latissimus dorsi and inserts on the olecranon. It acts to extend the elbow. It is innervated by the radial nerve.

Triceps brachii: Originates on the caudal border of the scapula and inserts on the olecranon tuber. It acts to extend the elbow and flex the shoulder. It is innervated by the radial nerve.

Anconeus: Originates on the humerus and inserts on the proximal end of the ulna. It acts to extend the elbow. It is innervated by the radial nerve.

Cranial muscles of the arm:

Biceps brachia: Originates on the supraglenoid tubercle and inserts on the ulnar and radial tuberosities. It acts to flex the elbow and extend the shoulder. It is innervated by the musculocutaneous nerve.

Brachialis: Originates on the lateral surface of humerus and inserts on the ulnar and radial tuberosities. It acts to flex the elbow. It is innervated by the musculocutaneous nerve.

Cranial and lateral muscles of antebrachium:

Extensor carpi radial: Originates on the supracondylar crest and inserts on the metacarpals. It acts to extend the carpus. It is innervated by the radial nerve.

Common digital extensor: Originates on the lateral epicondyle of the humerus and inserts on the distal phalanges. It acts to extend the carpus and joints of the digits 3, 4, and 5. It is innervated by the radial nerve.

Extensor carpi ulnar: Originates on the lateral epicondyle of the humerus and inserts on the metacarpal 5 and the accessory carpal bone. It acts to abduct and extend the carpal joint. It is innervated by the radial nerve.

Supinator: Originates on the lateral epicondyle of the humerus and inserts on the radius. It acts to rotate the forearm laterally. It is innervated by the radial nerve.

Abductor pollicis longus: Originates on the ulna and inserts on metacarpal 1. It acts to abduct the digit and extend the carpal joints. It is innervated by the radial nerve.

Caudal and medial muscles of forearm:

Pronator teres: Originates on the medial epicondyle of the humerus and inserts on the medial border of the radius. It acts to rotate forearm medially and flex the elbow. It is innervated by the median nerve.

Flexor carpi radial: Originates on the medial epicondyle of the humerus and inserts on the palmar side of metacarpals 2 and 3. It acts to flex the carpus. It is innervated by the median nerve.

Superficial digital flexor: Originates on the medial epicondyle of the humerus and inserts on the palmar surface of the middle phalanges. It acts to flex the carpus, metacarpophalangeal and proximal interphalangeal joints of the digits. It is innervated by the median nerve.

Flexor carpi ulnar: Originates on the olecranon and inserts on the accessory carpal bone. It acts to flex the carpus. It is innervated by the ulnar nerve.

Deep digital flexor: Originates on the medial epicondyle of the humerus and inserts on the palmar surface of the distal phalanx. It acts to flex the carpus, metacarpophalangeal joints, and the proximal and distal interphalangeal joints of the digits. It is innervated by the median nerve.

Pronator quadratus: Originates on surfaces of the radius and ulna. It acts to pronate the paw. It is innervated by the median nerve.

Caudal muscles of the thigh:

Biceps femoris: Originates on the ischiatic tuberosity and inserts on the patellar ligament. It acts to extend the hip, stifle and hock. It is innervated by the sciatic nerve.

Semitendinosus: Originates on the ischiatic tuberosity and inserts on the tibia. It acts to extend the hip, flex the stifle and extend the hock. It is innervated by the sciatic nerve.

Semimembranosus: Originates on the ischiatic tuberosity and inserts on the femur and tibia. It acts to extend the hip and stifle. It is innervated by the sciatic nerve.

Medial muscles of the thigh:

Sartorius: Originates on the ilium and inserts on the patella and tibia. It acts to flex the hip and both flex and extend the stifle. It is innervated by the femoral nerve.

Gracilis: Originates on the pelvic symphysis and inserts on the cranial border of the tibia. It acts to adduct the limb, flex the stifle and extend the hip and hock. It is innervated by the obturator nerve.

Pectineus: Originates on the iliopubic eminence and inserts on the caudal femur. It acts to adduct the limb. It is innervated by the obturator nerve.

Adductor: Originates on the pelvic symphysis and inserts on the lateral femur. It acts to adduct the limb and extend the hip. It is innervated by the obturator nerve.

Lateral muscles of the pelvis:

Tensor fasciae latae: Originates on the tuber coxae of the ilium and inserts on the lateral femoral fascia. It acts to flex the hip and extend the stifle. It is innervated by the cranial gluteal nerve.

Superficial gluteal: Originates on the lateral border of the sacrum and inserts on the 3rd trochanter. It acts to extend the hip and abduct the limb. It is innervated by the caudal gluteal nerve.

Middle gluteal: Originates on the ilium and inserts on the greater trochanter. It acts to abduct the hip and rotate the pelvic limb medially. It is innervated by the cranial gluteal nerve.

Deep gluteal: Originates on the ischiatic spine and inserts on the greater trochanter. It acts to extend the hip and rotate the pelvic limb medially. It is innervated by the cranial gluteal nerve.

Caudal hip muscles:

Internal obturator: Originates on the pelvic symphysis and inserts on the trochanteric fossa of the femur. It acts to rotate the pelvic limb laterally. It is innervated by the sciatic nerve.

Gemelli: Originates on the lateral surface of the ischium and inserts on the trochanteric fossa. It acts to rotate the pelvic limb laterally. It is innervated by the sciatic nerve.

Quadratus femoris: Originates on the ischium and inserts on the intertrochanteric crest. It acts to extend the hip and rotate the pelvic limb laterally.

External obturator: Originates on the pubis and ischium and inserts on the trochanteric fossa. It acts to rotate the pelvic limb laterally. It is innervated by the obturator nerve.

Cranial muscles of the thigh:

Quadriceps femoris: Originates on the femur and the ilium and inserts on the tibial tuberosity. It acts to extend the stifle and to flex the hip. It is innervated by the femoral nerve.

Ilipsoas: Originates on the ilium and inserts on the lesser trochanter. It acts to flex the hip. It is innervated by the femoral nerve.

Craniolateral muscles of the leg:

Cranial tibial: Originates on tibia and inserts on the plantar surfaces of metatarsals 1 and 2. It acts to flex the tarsus and rotates the paw laterally. It is innervated by the peroneal nerve.

Long digital extensor: Originates from the extensor fossa of the femur and inserts on the extensor processes of the distal phalanges. It acts to extend the digits and flex the tarsus. It is innervated by the peroneal nerve.

Peroneus longus: Originates on both the tibia and fibula and inserts on the 4th tarsal bone and the plantar aspect of the metatarsals. It acts to flex the tarsus and rotate the paw medially. It is innervated by the peroneal nerve.

Caudal muscles of the leg:

Gastrocnemius: Originates on the supracondylar tuberosities of the femur and inserts on the tuber calcanei. It acts to extend the tarsus and flex the stifle. It is innervated by the tibial nerve.

Superficial digital flexor: Originates on the lateral supracondylar tuberosity of the femur and inserts on the tuber calcanei and bases of the middle phalanges. It acts to flex the stifle and extend the tarsus. It is innervated by the tibial nerve.

Deep digital flexor: Originates on the fibular and inserts on the plantar surface of the distal phalanges. It acts to flex the digits and extend the tarsus. It is innervated by the tibial nerve.

Popliteus: Originates on the lateral condyle of the femur and inserts on the tibia. It acts to rotate the leg medially. It is innervated by the tibial nerve.

Skeleton

Bones and their significant points for muscle attachment:

Scapula: Spine of the Scapula, Supraglenoid Tubercle, Glenoid Cavity, Acromion Process, Supraspinous Fossa, Infraspinous Fossa, Neck, Coracoid, Process, Subscapular Fossa.

Humerus: Head of Humerus, Greater Tubercle, Lesser Tubercle, Intertubercular Groove, Deltopectoral Crest, Deltoid Tuberosity, Body of the Humerus, Epicondyles (Medial and Lateral), Humeral condyle (Trochlea and Capitulum, Radial and Olecranon Fossae).

Ulna and Radius: Olecranon Process, Trochlear Notch, Anconeal Process, Coronoid Processes (Medial and Lateral), Body of Ulna, Head of Radius, Body of Radius, Distal Trochlea, Styiloid Process (Medial and Lateral), Interosseus Space.

Metacarpals: Carpal Bones (Radial and Ulnar), Accessory Carpal Bone, First, Second, Third, and Fourth Metacarpals, Phalanges, Proximal Base, Body, Head, Ungual crest, Ungual process (Nails), Extensor process, Carpometacarpal Joints, Metacarpophalangeal Joints, Proximalinterphalangeal Joints, Interphalangeal Joints.

Femur: Head, Ligament of Head, Neck, Greater Trochanter, Lesser Trochanter, Trochanteric Fossa, Acetabulum Fossa (on Hip Bone), Distal Femur, Trochlea (and Ridges), Condyles (Medial/Lateral), Epicondyles (Medial/Lateral), Intercondylar Fossa, Extensor Fossa (Tiny Dent), Infrapatellar Fat Pad, Fabellae (Medial/Lateral).

Patella Kneecap

Tibia and Fibula: Tibial Condyles (Medial/Lateral), Intercondylar Eminences, Extensor Notch (Lateral), Tibial Tuberosity (Cranial), Tibial Cochlea, Medial Malleolus, Lateral Malleolus, Head of Fibula.

Metatarsals: Talus, Calcaneus, Trochlear Ridges, Central Tarsal Bone, First, Second, and Third Tarsal Bones

Vertebra Body, Pedicles, Laminae, Spinous Process, Transverse Process (Wings), Articular Process, Vertebral Foramen, Intervertebral Foramina, Atlas (C1), Axis (C2), dens, Ventral Lamina (on C6)

Pelvis: Acetabulum, Ilium, Ischium, Pubis

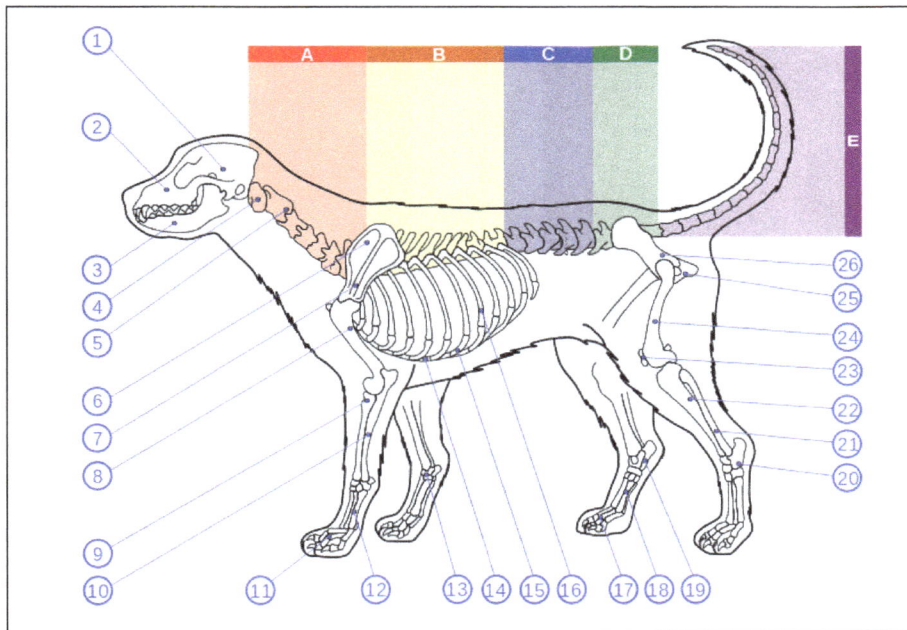

Skeleton of a dog. 1. Cranium 2. Maxilla 3. Mandible 4. Atlas 5. Axis 6. Scapula 7. Spine of scapula 8. Humerus 9. Radius 10. Ulna 11. Phalanges 12. Metacarpal Bones 13. Carpal Bones 14. Sternum 15. Cartilaginous part of rib 16. Ribs 17. Phalanges 18. Metatarsal Bones 19. Tarsal Bones 20. Calcaneus 21. Fibula 22. Tibia 23. Patella 24. Femur 25. Ischium 26. Pelvis.

Dog Skeletal Features

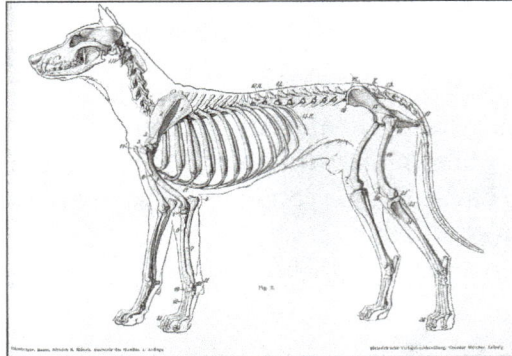

Lateral view of a dog skeleton.

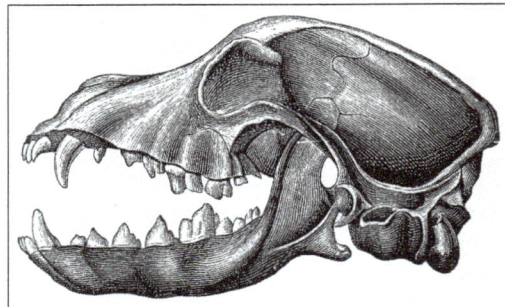

Lateral view of a dog skull - jaws open.

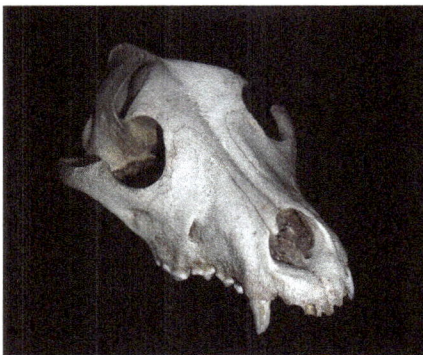

Frontal view of a dog skull.

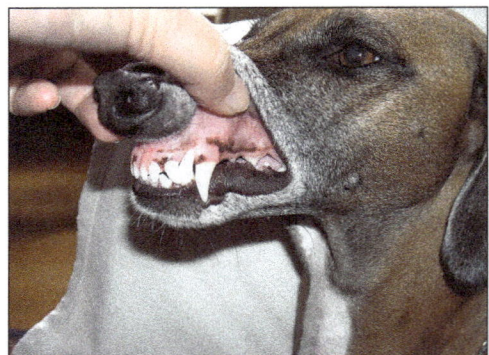

Image of dog teeth.

A study of skull morphology found that the domestic dog is morphologically distinct from all other canids except the wolf-like canids. "The difference in size and proportion between some breeds are as great as those between any wild genera, but all dogs are clearly members of the same species." In 2010, a study of dog skull shape compared to extant carnivorans proposed that "The greatest shape distances between dog breeds clearly surpass the maximum divergence between species in the Carnivora. Moreover, domestic dogs occupy a range of novel shapes outside the domain of wild carnivorans."

The domestic dog compared to the wolf shows the greatest variation in the size and shape of the skull that range from 7 to 28 cm in length. Wolves are dolichocephalic (long skulled) but not as extreme as some breeds of dogs such as greyhounds and Russian wolfhounds. Canine brachycephaly (short-skulledness) is found only in domestic dogs and is related to paedomorphosis. Puppies are born with short snouts, with the longer skull of dolichocephalic dogs emerging in later development. Other differences in head shape between brachycephalic and dolichocephalic dogs include changes in the craniofacial angle (angle between the basilar axis and hard palate), morphology of the temporomandibular joint, and radiographic anatomy of the cribriform plate.

One study found that the relative reduction in dog skull length compared to its width (the Cephalic Index) was significantly correlated to both the position and the angle of the brain within the skull. This was regardless of the brain size or the body weight of the dog.

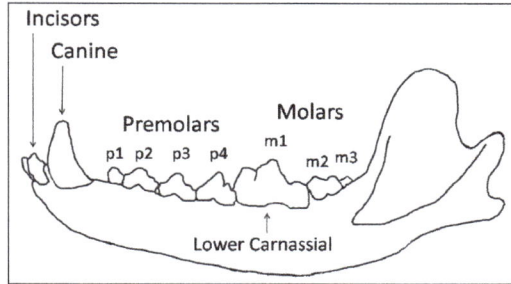

Wolf mandible diagram showing the names and positions of the teeth.

Bite force adjusted for body weight in Newtons per kilogram.

Canid	Carnassial	Canine
Wolf	131.6	127.3
Dhole	130.7	132.0
African wild dog	127.7	131.1
Greenland dog (domesticated)	117.4	114.3
Coyote	107.2	98.9
Side-striped jackal	93.0	87.5
Golden jackal	89.6	87.7
Black-backed jackal	80.6	78.3

Respiratory System

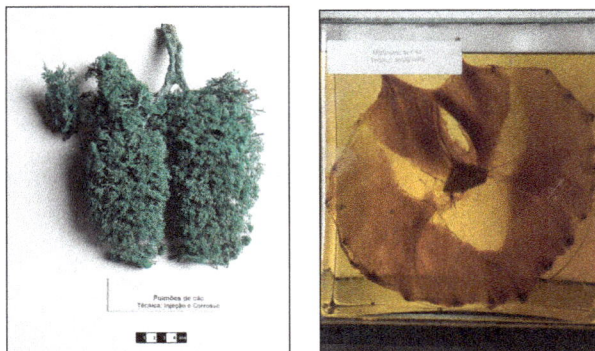

Dog Lungs (left) and Diaphragm (right).

The respiratory system is the set of organs responsible for the gas exchange between the animal's organism and the environment, that is, the pulmonary hematosis, allowing the cellular respiration.

This system has the main function to absorb oxygen and to eliminate much of the residual gases of the cells of the organism, like for example the carbon dioxide. As dogs have few sweat glands on their skin, this would explain the fact that they do not sweat, so the respiratory system also plays an important role in body thermoregulation.

Dogs are mammals with two large lungs and lobes, with a spongy appearance due to the presence of a system of delicate branches of the bronchioles in each lung, ending in closed, thin-walled chambers (the points of gas exchange) called alveoli.

The presence of a muscular structure, the diaphragm, exclusive of the mammals, divides the peritoneal cavity of the pleural cavity, besides assisting the ribs in the inhale.

Digestive System

The organs that make up the canine digestive system are:

- Mouth,
- Tongue,
- Esophagus,
- Stomach,
- Liver,
- Pancreas,
- Large intestine,
- Small intestine,
- Rectum,
- Anus.

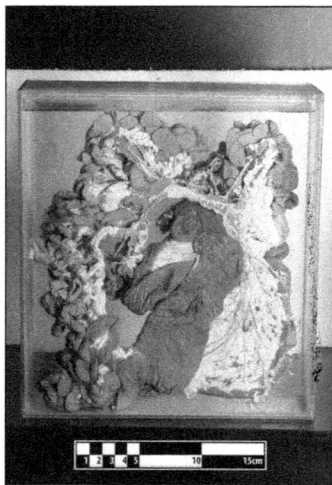

Dog digestive tract.

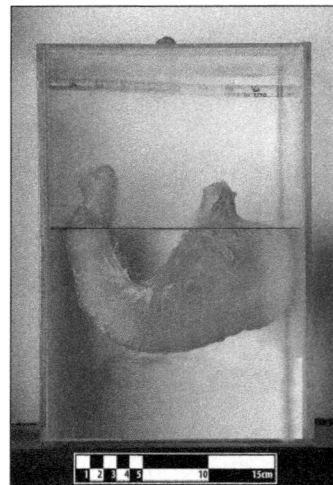

Dog stomach.

Vascular structure of the dog liver.

Reproductive System

Physical Characteristics

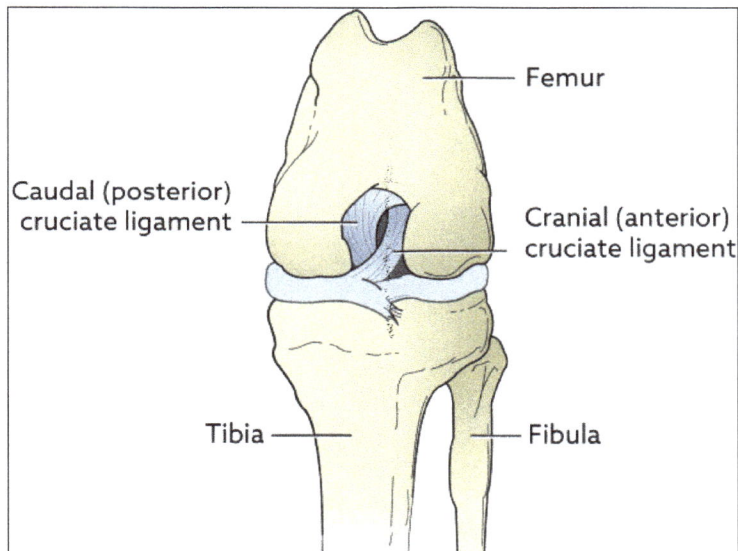

Dog knee.

Like most predatory mammals, the dog has powerful muscles, a cardiovascular system that supports both sprinting and endurance and teeth for catching, holding, and tearing.

The dog's skeleton provides the ability to jump and leap. Their legs can propel them forward rapidly, leaping as necessary to chase and overcome prey. They have small, tight feet, walking on their toes (thus having a digitigrade stance and locomotion). Their rear legs are fairly rigid and sturdy. The front legs are loose and flexible with only muscle attaching them to the torso.

The dog's muzzle size will come with the breed. The sizes of the muzzle have different names. Dogs with medium muzzles, such as the German Shepherd Dog, are called mesocephalic and dogs with a pushed in muzzle, such as the Pug, are called brachycephalic. Today's toy breeds have skeletons that mature in only a few months, while giant breeds, such as the Mastiffs, take 16 to 18 months for the skeleton to mature. Dwarfism has affected the proportions of some breeds' skeletons, as in the Basset Hound.

All dogs (and all living Canidae) have a ligament connecting the spinous process of their first thoracic (or chest) vertebra to the back of the axis bone (second cervical or neck bone), which supports the weight of the head without active muscle exertion, thus saving energy. This ligament is analogous in function (but different in exact structural detail) to the nuchal ligament found in ungulates. This ligament allows dogs to carry their heads while running long distances, such as while following scent trails with their nose to the ground, without expending much energy.

Dogs have disconnected shoulder bones (lacking the collar bone of the human skeleton) that allow a greater stride length for running and leaping. They walk on four toes, front and back, and have vestigial dewclaws on their front legs and on their rear legs. When a dog has extra dewclaws in addition to the usual one in the rear, the dog is said to be "double dewclawed."

Size

The difference in overall body size between a Cane Corso (Italian mastiff) and a Yorkshire terrier is over 30-fold, yet both are members of the same species.

Dogs are highly variable in height and weight. The smallest known adult dog was a Yorkshire Terrier that stood only 6.3 cm (2.5 in) at the shoulder, 9.5 cm (3.7 in) in length along the head and body, and weighed only 113 grams (4.0 oz). The largest known adult dog was an English Mastiff which weighed 155.6 kg (343 lb) and was 250 cm (98 in) from the snout to the tail. The tallest known adult dog is a Great Dane that stands 106.7 cm (42.0 in) at the shoulder.

In 2007, a study identified a gene that is proposed as being responsible for size. The study found a regulatory sequence next to the gene Insulin-like growth factor 1 (IGF1) and together with the gene and regulatory sequence is a major contributor to body size in all small dogs. Two variants of this gene were found in large dogs, making a more complex reason for large breed size. The researchers concluded this gene's instructions to make dogs small must be at least 12,000 years old and it is not found in wolves. Another study has proposed that lap dogs (small dogs) are among the oldest existing dog types.

Coat

Montage showing the coat variation of the dog.

Domestic dogs often display the remnants of countershading, a common natural camouflage pattern. The general theory of countershading is that an animal that is lit from above will appear lighter on its upper half and darker on its lower half where it will usually be in its own shade. This is a pattern that predators can learn to watch for. A counter shaded animal will have dark coloring on its upper surfaces and light coloring below. This reduces the general visibility of the animal. One reminder of this pattern is that many breeds will have the occasional "blaze", stripe, or "star" of white fur on their chest or undersides.

A study found that the genetic basis that explains coat colors in horse coats and cat coats did not apply to dog coats. The project took samples from 38 different breeds to find the gene (a beta defensin gene) responsible for dog coat color. One version produces yellow dogs and a mutation produces black. All dog coat colors are modifications of black or yellow. For example, the white in white miniature schnauzers is a cream color, not albinism (a genotype of e/e at MC1R).

Modern dog breeds exhibit a diverse array of fur coats, including dogs without fur, such as the Mexican Hairless Dog. Dog coats vary in texture, color, and markings, and a specialized vocabulary has evolved to describe each characteristic.

Tail

There are many different shapes of dog tails: straight, straight up, sickle, curled and cork-screw. In some breeds, the tail is traditionally docked to avoid injuries (especially for hunting dogs). It can happen that some puppies are born with a short tail or no tail in some breeds. Dogs have a violet gland or supracaudal gland on the dorsal (upper) surface of their tails.

Footpad

Dogs can stand, walk and run on snow and ice for long periods of time. When a dog's footpad is exposed to the cold, heat loss is prevented by an adaptation of the blood system that recirculates heat back into the body. It brings blood from the skin surface and retains warm blood in the pad surface.

Dewclaw

There is some debate about whether a dewclaw helps dogs to gain traction when they run because, in some dogs, the dewclaw makes contact when they are running and the nail on the dewclaw often wears down in the same way that the nails on their other toes do from contact with the ground. However, in many dogs, the dewclaws never make contact with the ground. In this case, the dewclaw's nail never wears away and it is then often trimmed to keep it to a safe length.

The dewclaws are not dead appendages. They can be used to lightly grip bones and other items that dogs hold with their paws. However, in some dogs, these claws may not appear to be connected to the leg at all except by a flap of skin. In such dogs, the claws do not have a use for gripping as the claw can easily fold or turn.

There is also some debate as to whether dewclaws should be surgically removed. The argument for removal states that dewclaws are a weak digit, barely attached to the leg, so they can rip partially off or easily catch on something and break which can be extremely painful and prone to infection. Others say the pain of removing a dewclaw is far greater than any other risk. For this reason, removal of dewclaws is illegal in many countries. There is, perhaps, an exception for hunting dogs who can sometimes tear the dewclaw while running in overgrown vegetation. If a dewclaw is to be removed, this should be done when the dog is a puppy, sometimes as young as 3 days old; although, it can also be performed on older dogs if necessary (the surgery may be more difficult then). The surgery is fairly straightforward and may even be done with only local anesthetics if the digit is not well connected to the leg. Unfortunately, many dogs can't resist licking at their sore paws following the surgery, so owners need to remain vigilant in their aftercare.

In addition, for those dogs whose dewclaws make contact with the ground when they run, it is possible that removing them could be a disadvantage for a dog's speed in running and changing direction, particularly in performance dog sports such as dog agility.

Senses

Vision

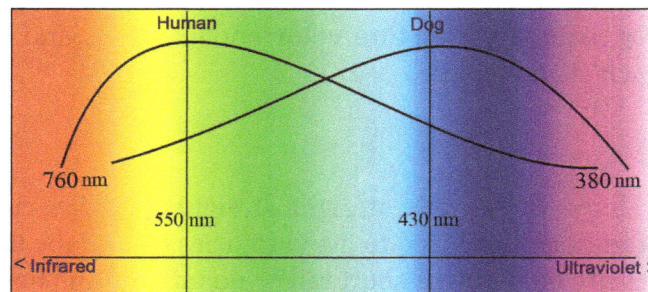

Frequency sensitivity compared with humans.

Like most mammals, dogs have only two types of cone photoreceptor, making them dichromats. These cone cells are maximally sensitive between 429 nm and 555 nm. Behavioural studies have shown that the dog's visual world consists of yellows, blues and grays, but they have difficulty differentiating red and green making their color vision equivalent to red–green color blindness in humans (deuteranopia). When a human perceives an object as "red," this object appears as "yellow" to the dog and the human perception of "green" appears as "white," a shade of gray. This white region (the neutral point) occurs around 480 nm, the part of the spectrum which appears blue-green to humans. For dogs, wavelengths longer than the neutral point cannot be distinguished from each other and all appear as yellow.

Dogs use color instead of brightness to differentiate light or dark blue/yellow. They are less sensitive to differences in grey shades than humans and also can detect brightness at about half the accuracy of humans.

The dog's visual system has evolved to aid proficient hunting. While a dog's visual acuity is poor (that of a poodle's has been estimated to translate to a Snellen rating of 20/75), their visual discrimination for moving objects is very high. Dogs have been shown to be able to discriminate between humans (e.g. identifying their human guardian) at a range of between 800 and 900 metres (2,600 and 3,000 ft); however, this range decreases to 500–600 metres (1,600–2,000 ft) if the object is stationary.

Dogs can detect a change in movement that exists in a single diopter of space within their eye. Humans, by comparison, require a change of between 10 and 20 diopters to detect movement.

As crepuscular hunters, dogs often rely on their vision in low light situations. They have very large pupils, a high density of rods in the fovea, an increased flicker rate, and a tapetum lucidum. The tapetum is a reflective surface behind the retina that reflects light to give the photoreceptors a second chance to catch the photons. There is also a relationship between body size and overall diameter of the eye. A range of 9.5 and 11.6 mm can be found between various breeds of dogs. This 20% variance can be substantial and is associated as an adaptation toward superior night vision.

The eyes of different breeds of dogs have different shapes, dimensions, and retina configurations. Many long-nosed breeds have a "visual streak"—a wide foveal region that runs across the width of the retina and gives them a very wide field of excellent vision. Some long-muzzled breeds, in particular, the sighthounds, have a field of vision up to 270° (compared to 180° for humans). Short-nosed breeds, on the other hand, have an "area centralis": a central patch with up to three times the density of nerve endings as the visual streak, giving them detailed sight much more like a human's. Some broad-headed breeds with short noses have a field of vision similar to that of humans.

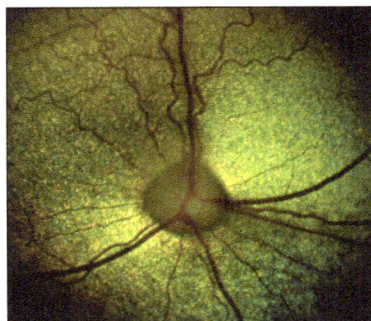

Dog retina showing optic disc and vasculature.

Most breeds have good vision, but some show a genetic predisposition for myopia such as Rott-weilers, with which one out of every two has been found to be myopic. Dogs also have a greater divergence of the eye axis than humans, enabling them to rotate their pupils farther in any direction. The divergence of the eye axis of dogs ranges from 12–25° depending on the breed.

Experimentation has proven that dogs can distinguish between complex visual images such as that of a cube or a prism. Dogs also show attraction to static visual images such as the silhouette of a dog on a screen, their own reflections, or videos of dogs; however, their interest declines sharply once they are unable to make social contact with the image.

Hearing

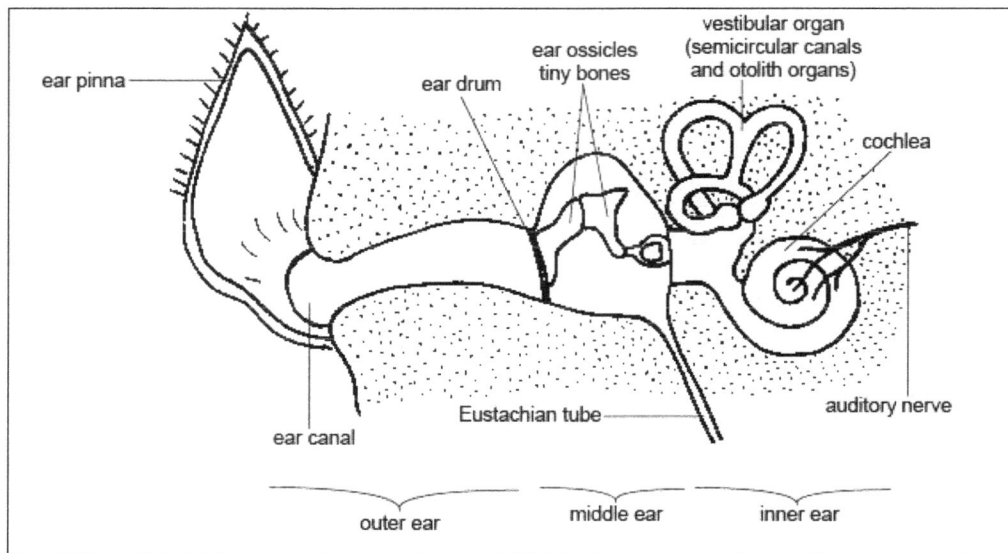

Anatomy of the ear.

The frequency range of dog hearing is between 16-40 Hz (compared to 20–70 Hz for humans) and up to 45–60 kHz (compared to 13–20 kHz for humans), which means that dogs can detect sounds far beyond the upper limit of the human auditory spectrum.

Dogs have ear mobility that allows them to rapidly pinpoint the exact location of a sound. Eighteen or more muscles can tilt, rotate, raise, or lower a dog's ear. A dog can identify a sound's location much faster than a human can, as well as hear sounds at four times the distance.

Those with more natural ear shapes, like those of wild canids like the fox, generally hear better than those with the floppier ears of many domesticated species.

Smell

While the human brain is dominated by a large visual cortex, the dog brain is dominated by a large olfactory cortex. Dogs have roughly forty times more smell-sensitive receptors than humans, ranging from about 125 million to nearly 300 million in some dog breeds, such as bloodhounds. This is thought to make its sense of smell up to 40 times more sensitive than human's. These receptors are spread over an area about the size of a pocket handkerchief (compared to 5 million over an area the size of a postage stamp for humans). Dogs' sense of smell also includes the use of the vomeronasal organ, which is used primarily for social interactions.

The dog has mobile nostrils that help it determine the direction of the scent. Unlike humans, the dog does not need to fill up his lungs as he continuously brings the odor into his nose in bursts of 3-7 sniffs. The dog's nose has a bony structure inside that humans don't have, which allows the air that has been sniffed to pass over a bony shelf and many odor molecules stick to it. The air above this shelf is not washed out when the dog breathes normally, so the scent molecules accumulate in the nasal chambers and the scent builds with intensity, allowing the dog to detect the faintest of odors.

One study into the learning ability of dogs compared to wolves indicated that dogs have a better sense of smell than wolves when locating hidden food, but there has yet been no experimental data to support this view.

The wet nose, or rhinarium, is essential for determining the direction of the air current containing the smell. Cold receptors in the skin are sensitive to the cooling of the skin by evaporation of the moisture by air currents.

Scent hounds, especially the Bloodhound, are bred for their keen sense of smell.

Taste

Dogs have around 1,700 taste buds compared to humans with around 9,000. The sweet taste buds in dogs respond to a chemical called furaneol which is found in many fruits and in tomatoes. It appears that dogs do like this flavor and it probably evolved because in a natural environment dogs frequently supplement their diet of small animals with whatever fruits happen to be available. Because of dogs' dislike of bitter tastes, various sprays, and gels have been designed to keep dogs from chewing on furniture or other objects. Dogs also have taste buds that are tuned for water, which is something they share with other carnivores but is not found in humans. This taste sense is found at the tip of the dog's tongue, which is the part of the tongue that he curls to lap water. This area responds to water at all times, but when the dog has eaten salty or sugary foods the sensitivity to the taste of water increases. It is proposed that this ability to taste water evolved as a way for the body to keep internal fluids in balance after the animal has eaten things that will either result in more urine being passed or will require more water to adequately process. It certainly appears that when these special water taste buds are active, dogs seem to get an extra pleasure out of drinking water, and will drink copious amounts of it.

Touch

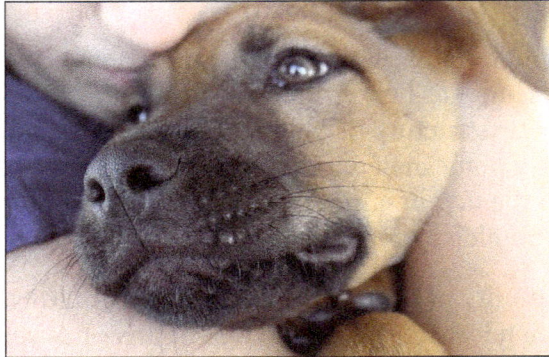
A dog's whiskers act as sensing organs.

The main difference between human and dog touch is the presence of specialized whiskers known as vibrissae. Vibrissae are present above the dog's eyes, below their jaw, and on their muzzle. They are sophisticated sensing organs. Vibrissae are more rigid and embedded much more deeply in the skin than other hairs and have a greater number of receptor cells at their base. They can detect air currents, subtle vibrations, and objects in the dark. They provide an early warning system for objects that might strike the face or eyes, and probably help direct food and objects towards the mouth.

Magnetic Sensitivity

Dogs may prefer, when they are off the leash and Earth's magnetic field is calm, to urinate and defecate with their bodies aligned on a north-south axis. Another study suggested that dogs can see the earth's magnetic field.

Temperature Regulation

The highly sensitive nose of a dog.

Primarily, dogs regulate their body temperature through panting and sweating via their paws. Panting moves cooling air over the moist surfaces of the tongue and lungs, transferring heat to the atmosphere.

Dogs and other canids also possess a very well-developed set of nasal turbinates, an elaborate set of bones and associated soft-tissue structures (including arteries and veins) in the nasal cavities. These turbinates allow for heat exchange between small arteries and veins on their maxilloturbinate surfaces (the surfaces of turbinates positioned on maxilla bone) in a counter-current heat-exchange system. Dogs are capable of prolonged chases, in contrast to the ambush predation of cats, and these complex turbinates play an important role in enabling this (cats only possess a much smaller and less-developed set of nasal turbinates). This same complex turbinate structure helps conserve water in arid environments. The water conservation and thermoregulatory capabilities of these well-developed turbinates in dogs may have been crucial adaptations that allowed dogs (including both domestic dogs and their wild prehistoric ancestors) to survive in the harsh Arctic environment and other cold areas of northern Eurasia and North America, which are both very dry and very cold.

Felidae Anatomy and Physiology

Cat Anatomy and Physiology

The anatomy of the domestic cat is similar to that of other members of the genus Felis.

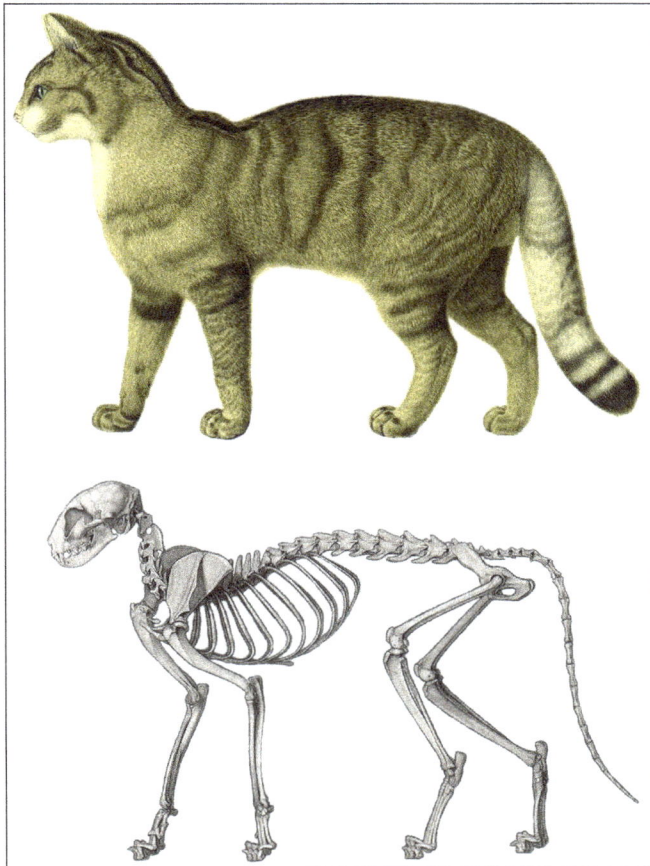

Skeleton of a domestic cat.

Mouth

A yawning cat, exposing its mouth.

Cats have highly specialized teeth for killing prey and tearing meat. The premolar and first molar, together called the carnassial pair, are located on each side of the mouth. These teeth efficiently function to shear meat like a pair of scissors. While this feature is present in canids, it is highly developed in felines.

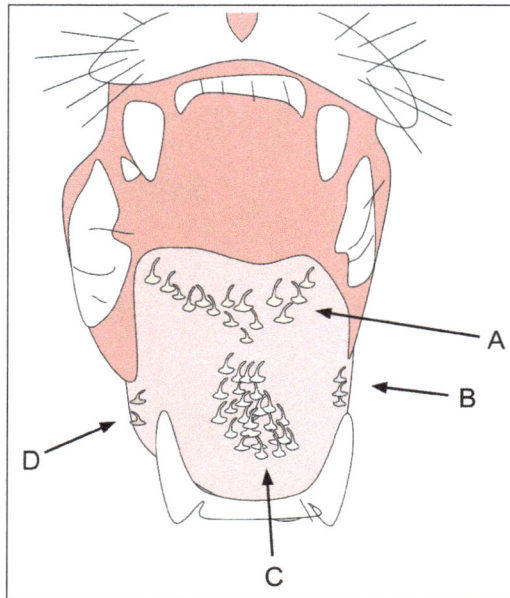

A cat tongue holds 4 different types of papillae. Arrows B & D point to the papillae used for taste. While arrow C Is pointing to the filiform papillae which assist in grooming & removing flesh from prey. (A) points to the circumvallate papillae which assist with taste.

The cat's tongue has sharp spines, or papillae, useful for retaining and ripping flesh from a carcass. These papillae are small backward-facing hooks that contain keratin, and also assist in their grooming. The papillae also help hold water on the tongue while drinking.

The cat's oral structures provide for a variety of vocalizations used for communication, including meowing, purring, hissing, growling, squeaking, chirping, clicking, and grunting.

Ears

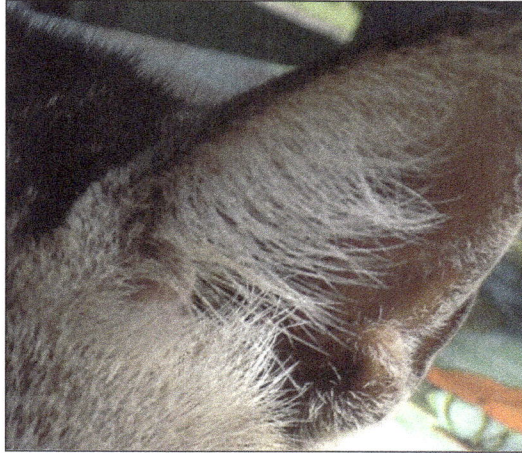

A cat's ear which has special fur for sensing and protection.

Thirty-two individual muscles in each ear allow for a kind of directional hearing; a cat can move each ear independently of the other. Because of this mobility, a cat can move its body in one direction and point its ears in another direction. Most cats have straight ears pointing upward. Unlike with dogs, flap-eared breeds are extremely rare (*Scottish Folds* have one such exceptional mutation). When angry or frightened, a cat will lay back its ears to accompany the growling or hissing sounds it makes. Cats also turn their ears back when they are playing or to listen to a sound coming from behind them. The fold of skin forming a pouch on the lower posterior part of the ear, known as Henry's pocket, is usually prominent in a cat's ear. Its function is unknown, though it may assist in filtering sounds.

Nose

A cat's nose is highly adapted.

Cats are highly territorial, and secretion of odors plays a major role in cat communication. The nose helps cats to identify territories, other cats and mates, to locate food, and has various other uses. A cat's sense of smell is believed to be about fourteen times stronger than that of humans.

The rhinarium is quite tough, to allow it to absorb rather rough treatment sometimes. The color varies according to the genotype (genetic makeup) of the cat. A cat's skin has the same color as the fur, but the color of the nose leather is probably dictated by a dedicated gene. Cats with white fur have skin susceptible to damage by ultraviolet light, which may cause cancer. Extra care is required when outside in the hot sun.

Legs

Cats, like dogs, are digitigrades. They walk directly on their toes, with the bones of their feet making up the lower part of the visible leg. All cats are capable of walking very precisely. Like all felines, they directly register; that is, they place each hind paw almost directly in the print of the corresponding forepaw, minimizing noise and visible tracks. This also provides sure footing for their hind paws when they navigate rough terrain. The two back legs allow falling and leaping far distances without injury.

Unlike most mammals, when cats walk, they use a "pacing" gait; that is, they move the two legs on one side of the body before the legs on the other side. This trait is shared with camels and giraffes. As a walk speeds up into a trot, a cat's gait will change to be a "diagonal" gait, similar to that of most other mammals: the diagonally opposite hind and forelegs will move simultaneously. Cat height can vary depending on breed and gender, but is usually around 12 inches or 30.5 centimeters.

Claws

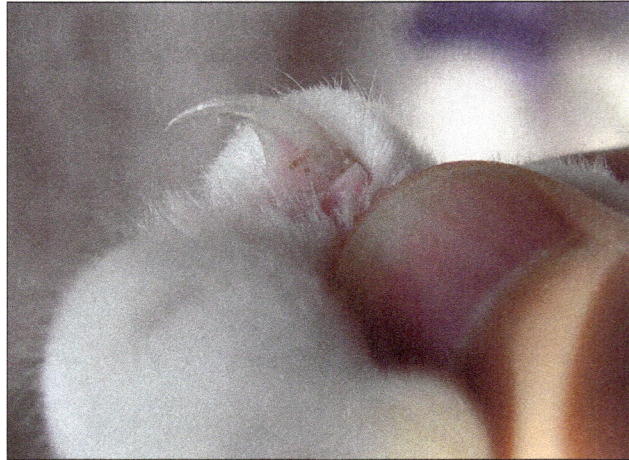

A cat's claw.

Like nearly all members of the family Felidae, cats have protractable claws. In their normal, relaxed position, the claws are sheathed with the skin and fur around the toe pads. This keeps the claws sharp by preventing wear from contact with the ground and allows the silent stalking of prey. The claws on the forefeet are typically sharper than those on the hind feet. Cats can voluntarily extend their claws on one or more paws. They may extend their claws in hunting or self-defense, climbing, "kneading", or for extra traction on soft surfaces (bedspreads, thick rugs, skin, etc.). It is also possible to make a cooperative cat extend its claws by carefully pressing both the top and bottom of the paw. The curved claws can become entangled in carpet or thick fabric, which can cause injury if the cat is unable to free itself.

Most cats have a total of 18 digits and claws. 5 on each forefoot, the 5th digit being the dewclaw; and 4 on each hind foot. The dewclaw is located high on the foreleg, is not in contact with the ground and is non-weight bearing.

Some cats can have more than 18 digits, due to a common mutation called polydactyly or polydactylism, which can result in five to seven toes per paw.

Temperature and Heart Rate

Two cats sharing body heat.

The normal body temperature of a cat is between 38.3 and 39.0 °C (100.9 and 102.2 °F). A cat is considered *febrile* (hyperthermic) if it has a temperature of 39.5 °C (103.1 °F) or greater, or hypothermic if less than 37.5 °C (99.5 °F). For comparison, humans have an average body temperature of about 37.0 °C (98.6 °F). A domestic cat's normal heart rate ranges from 140 to 220 beats per minute (bpm), and is largely dependent on how excited the cat is. For a cat at rest, the average heart rate usually is between 150 and 180 bpm, more than twice that of a human, which averages 70 bpm.

Skin

Cats possess rather loose skin; this allows them to turn and confront a predator or another cat in a fight, even when it has a grip on them. This is also an advantage for veterinary purposes, as it simplifies injections. In fact, the lives of cats with kidney failure can sometimes be extended for years by the regular injection of large volumes of fluid subcutaneously, which serves as an alternative to dialysis.

Scruff

The particularly loose skin at the back of the neck is known as the scruff, and is the area by which a mother cat grips her kittens to carry them. As a result, cats tend to become quiet and passive when gripped there. This behavior also extends into adulthood, when a male will grab the female by the scruff to immobilize her while he mounts, and to prevent her from running away as the mating process takes place.

This technique can be useful when attempting to treat or move an uncooperative cat. However, since an adult cat is heavier than a kitten, a pet cat should never be carried by the scruff, but should instead have its weight supported at the rump and hind legs, and at the chest and front paws.

Primordial Pouches

Some cats share common traits due to heredity. One of those is the primordial pouch, sometimes referred to as "spay sway" by owners who notice it once the cat has been spayed or neutered. It is located on a cat's belly. Its appearance is similar to a loose flap of skin that might occur if the cat had been overweight and had then lost weight. It provides a little extra protection against kicks, which are common during cat fights as a cat will try to rake with its rear claws. In wild cats, the ancestors of domesticated felines, this pouch appears to be present to provide extra room in case the animal has the opportunity to eat a large meal and the stomach needs to expand. This stomach pouch also allows the cat to bend and expand, allowing for faster running and higher jumping.

Skeleton

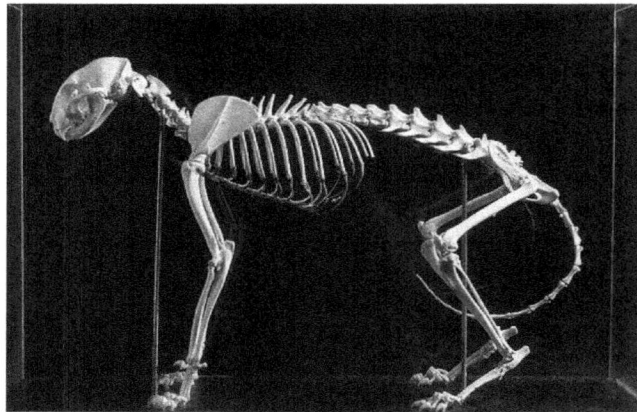

Cat skeleton.

Cats have seven cervical vertebrae like almost all mammals, thirteen thoracic vertebrae (humans have twelve), seven lumbar vertebrae (humans have five), three sacral vertebrae (humans have five because of their bipedal posture), and, except for Manx cats and other shorter tailed cats, twenty-two or twenty-three caudal vertebrae (humans have three to five, fused into an internal coccyx). The extra lumbar and thoracic vertebrae account for the cat's enhanced spinal mobility and flexibility, compared to humans. The caudal vertebrae form the tail, used by the cat as a counterbalance to the body during quick movements. Between their vertebrae, they have elastic discs, useful for cushioning the jump landings.

Unlike human arms, cat forelimbs are attached to the shoulder by free-floating clavicle bones, which allows them to pass their body through any space into which they can fit their heads.

Skull

The cat skull is unusual among mammals in having very large eye sockets and a powerful and specialized jaw. Compared to other felines, domestic cats have narrowly spaced canine teeth, adapted to their preferred prey of small rodents.

Muscles

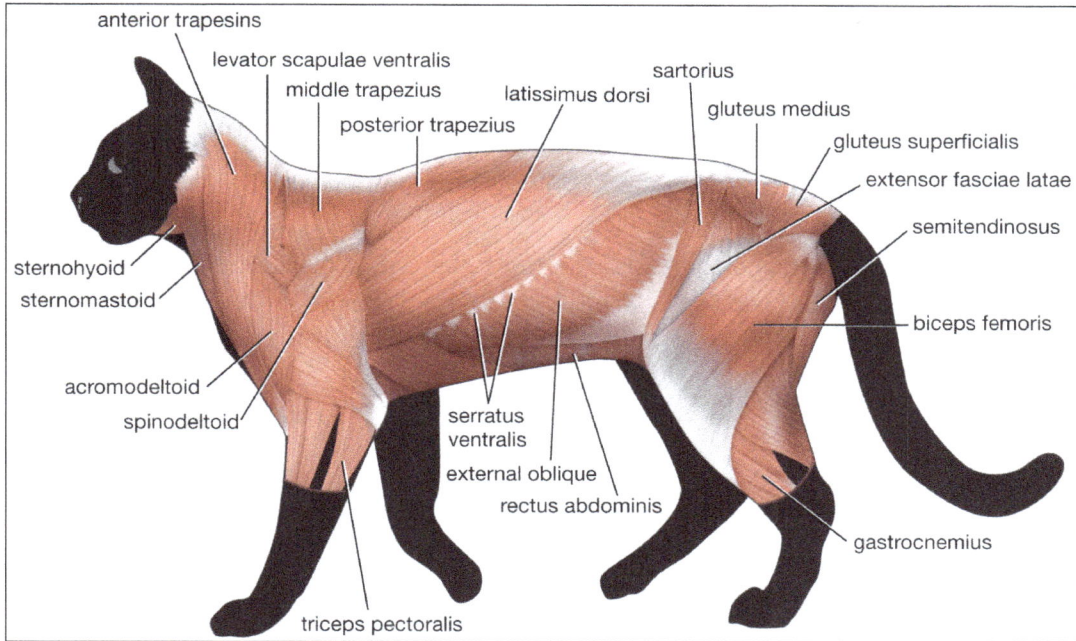

Diagram of the muscular system of a cat.

Labels (clockwise): anterior trapesins, levator scapulae ventralis, middle trapezius, posterior trapezius, latissimus dorsi, sartorius, gluteus medius, gluteus superficialis, extensor fasciae latae, semitendinosus, biceps femoris, gastrocnemius, rectus abdominis, external oblique, serratus ventralis, triceps pectoralis, spinodeltoid, acromodeltoid, sternomastoid, sternohyoid

Internal Abdominal Oblique

This muscle's origin is the lumbodorsal fascia and ribs. Its insertion is at the pubis and linea alba (via aponeurosis), and its action is the compression of abdominal contents. It also laterally flexes and rotates the vertebral column.

Transversus Abdominis

This muscle is the innermost abdominal muscle. Its origin is the second sheet of the lumbodorsal fascia and the pelvic girdle and its insertion is the linea alba. Its action is the compression of the abdomen.

Rectus Abdominis

To see this muscle, first remove the extensive aponeurosis situated on the ventral surface of the cat. Its fibers are extremely longitudinal, on each side of the linea alba. It is also traversed by the inscriptiones tendinae, or what others called *myosepta*.

Deltoid

The deltoid muscles lie just lateral to the trapezius muscles, originating from several fibers spanning the clavicle and scapula, converging to insert at the humerus. Anatomically, there are only two deltoids in the cat, the acromiodeltoid and the spinodeltoid. However, to conform to human anatomy standards, the clavobrachialis is now also considered a deltoid and is commonly referred to as the clavodeltoid.

Acromiodeltoid

The acromiodeltoid is the shortest of the deltoid muscles. It lies lateral to (to the side of) the clavodeltoid, and in a more husky cat it can only be seen by lifting or reflecting the clavodeltoid. It originates at the acromion process and inserts at the deltoid ridge. When contracted, it raises and rotates the humerus outward.

Spinodeltoid

A stout and short muscle lying posterior to the acromiodeltoid. It lies along the lower border of the scapula, and it passes through the upper arm, across the upper end of muscles of the upper arm. It originates at the spine of the scapula and inserts at the deltoid ridge. Its action is to raise and rotate the humerus outward.

Head

Masseter

The masseter is a great, powerful, and very thick muscle covered by a tough, shining fascia lying ventral to the zygomatic arch, which is its origin. It inserts into the posterior half of the lateral surface of the mandible. Its action is the elevation of the mandible (closing of the jaw).

Temporalis

The temporalis is a great mass of mandibular muscle, and is also covered by a tough and shiny fascia. It lies dorsal to the zygomatic arch and fills the temporal fossa of the skull. It arises from the side of the skull and inserts into the coronoid process of the mandible. It too, elevates the jaw.

Integumental

The two main integumentary muscles of a cat are the *platysma* and the *cutaneous maximus*. The *cutaneous maximus* covers the dorsal region of the cat and allows it to shake its skin. The *platysma* covers the neck and allows the cat to stretch the skin over the pectoralis major and deltoid muscles.

Neck and Back

Rhomboideus

The rhomboideus is a thick, large muscle below the trapezius muscles. It extends from the vertebral border of the scapula to the mid-dorsal line. Its origin is from the neural spines of the first four thoracic vertebrae, and its insertion is at the vertebral border of the scapula. Its action is to draw the scapula to the dorsal.

Rhomboideus Capitis

The rhomboideus capitis is the most cranial of the deeper muscles. It is underneath the clavotrapezius. Its origin is the superior nuchal line, and its insertion is at the scapula. Action draws scapula cranially.

Splenius

The splenius is the most superficial of all the deep muscles. It is a thin, broad sheet of muscle underneath the clavotrapezius and deflecting it. It is crossed also by the rhomboideus capitis. Its origin is the mid-dorsal line of the neck and fascia. The insertion is the superior nuchal line and atlas. It raises or turns the head.

Serratus Ventralis

The serratus ventralis is exposed by cutting the wing-like latissimus dorsi. The said muscle is covered entirely by adipose tissue. The origin is from the first nine or ten ribs and from part of the cervical vertebrae.

Serratus Dorsalis

The serratus dorsalis is medial to both the scapula and the serratus ventralis. Its origin is via apoeurosis following the length of the mid-dorsal line, and its insertion is the dorsal portion of the last ribs. Its action is to depress and retracts the ribs during breathing.

Intercostals

The intercostals are a set of muscles sandwiched among the ribs. They interconnect ribs, and are therefore the primary respiratory skeletal muscles. They are divided into the *external* and the *internal subscapularis*. The origin and insertion are in the ribs. The intercostals pull the ribs backwards or forwards.

Caudofemoralis

The caudofemoralis is a muscle found in the pelvic limb. The caudofemoralis acts to flex the tail laterally to its respective side when the pelvic limb is bearing weight. When the pelvic limb is lifted off the ground, contraction of the caudofemoralis causes the limb to abduct and the shank to extend by extending the hip joint.

Pectoral

Pectoantebrachialis

Pectoantebrachialis muscle is just one-half inch wide and is the most superficial in the pectoral muscles. Its origin is the manubrium of the sternum, and its insertion is in a flat tendon on the fascia of the proximal end of the ulna. Its action is to draw the arm towards the chest. There is no human equivalent.

Pectoralis Major

The pectoralis major, also called *pectoralis superficialis*, is a broad triangular portion of the pectoralis muscle which is immediately below the pectoantebrachialis. It is smaller than the pectoralis minor muscle. Its origin is the sternum and median ventral raphe, and its insertion is at the humerus. Its action is to draw the arm towards the chest.

Pectoralis Minor

The pectoralis minor muscle is larger than the pectoralis major. However, most of its anterior border is covered by the pectoralis major. Its origins are ribs three–five, and its insertion is the coracoid process of the scapula. Its actions are the tipping of the scapula and the elevation of ribs three–five.

Xiphihumeralis

The most posterior, flat, thin, and long strip of pectoral muscle is the xiphihumeralis. It is a band of parallel fibers that is found in felines but not in humans. Its origin is the xiphoid process of the sternum. The insertion is the humerus.

Trapezius

In the cat there are three thin flat muscles that cover the back, and to a lesser extent, the neck. They pull the scapula toward the mid-dorsal line, anteriorly, and posteriorly.

Clavotrapezius

The most anterior of the trapezius muscles, it is also the largest. Its fibers run obliquely to the ventral surface. Its origin is the superior nuchal line and median dorsal line and its insertion is the clavicle. Its action is to draw the clavicle dorsally and towards the head.

Acromiotrapezius

Acromiotrapezius is the middle trapezius muscle. It covers the dorsal and lateral surfaces of the scapula. Its origin is the neural spines of the cervical vertebrae and its insertion is in the metacromion process and fascia of the clavotrapezius. Its action is to draw the scapula to the dorsal, and hold the two scapula together.

Spinotrapezius

Spinotrapezius, also called *thoracic trapezius*, is the most posterior of the three. It is triangular shaped. Posterior to the acromiotrapezius and overlaps latissimus dorsi on the front. Its origin is the neural spines of the thoracic vertebrae and its insertion is the scapular fascia. Its action is to draw the scapula to the dorsal and caudal region.

Digestive System

The digestion system of cats begins with their sharp teeth and abrasive tongue papillae, which help them tear meat, which is most, if not all, of their diet. Cats naturally do not have a diet high in carbohydrates, and therefore, their saliva doesn't contain the enzyme amylase. Food moves from the mouth through the esophagus and into the stomach. The gastrointestinal tract of domestic cats contains a small cecum and unsacculated colon. The cecum while similar to dogs, doesn't have a coiled cecum.

The stomach of the cat can be divided into distinct regions of motor activity. The proximal end of the stomach relaxes when food is digested. While food is being digested this portion of the stomach either has rapid stationary contractions or a sustained tonic contraction of muscle. These different actions result in either the food being moved around or the food moving towards the distal portion of the stomach. The distal portion of the stomach undergoes rhythmic cycles of partial depolarization. This depolarization sensitizes muscle cells so they are more likely to contract. The stomach is not only a muscular structure, it also serves a chemical function by releasing hydrochloric acid and other digestive enzymes to break down food.

Food moves from the stomach into the small intestine. The first part of the small intestine is the duodenum. As food moves through the duodenum, it mixes with bile, a fluid that neutralizes stomach acid and emulsifies fat. The pancreas releases enzymes that aid in digestion so that nutrients can be broken down and pass through the intestinal mucosa into the blood and travel to the rest of the body. The pancreas doesn't produce starch processing enzymes because cats don't eat a diet high in carbohydrates. Since the cat digests low amounts of glucose, the pancreas uses amino acids to trigger insulin release instead.

Food then moves on to the jejunum. This is the most nutrient absorptive section of the small intestine. The liver regulates the level of nutrients absorbed into the blood system from the small intestine. From the jejunum, whatever food that has not been absorbed is sent to the ileum which connects to the large intestine. The first part of the large intestine is the cecum and the second portion is the colon. The large intestine reabsorbs water and forms fecal matter.

There are some things that the cats are not able to digest. For example, cats clean themselves by licking their fur with their tongue, which causes them to swallow a lot of fur. This causes a build-up of fur in a cat's stomach and creates a mass of fur. This is often thrown up and is better known as a hair ball.

The short length of the digestive tract of the cat causes cats' digestive system to weigh less than other species of animals, which allows cats to be active predators. While cats are well adapted to be predators they have a limited ability to regulate catabolic enzymes of amino acids meaning amino acids are constantly being destroyed and not absorbed. Therefore, cats require a higher protein proportion in their diet than many other species. Cats are not adapted to synthesize niacin from tryptophan and, because they are carnivores, can't convert carotene to vitamin A, so eating plants while not harmful does not provide them nutrients.

Genitalia

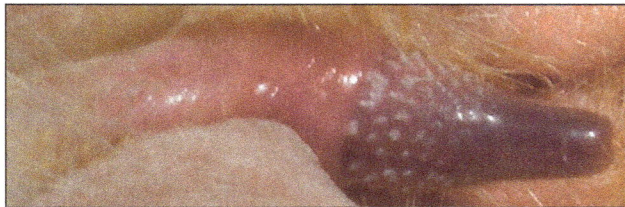

Penile spines of a domestic cat.

Female Genitalia

In the female cat, the genitalia includes the uterus, the vagina, the genital passages and teats. Together with the vulva, the vagina of the cat is involved in mating and provides a channel for newborns during parturition, or birth. The vagina is long and wide. Genital passages are the oviducts of the cat. They are short, narrow, and not very sinuous.

Male Genitalia

In the male cat, the genitalia includes two gonads and the penis, which is covered with small spines.

Physiology

Normal physiological values.

Body temperature	38.6 °C (101.5 °F)
Heart rate	120–140 beats per minute
Breathing rate	16–40 breaths per minute

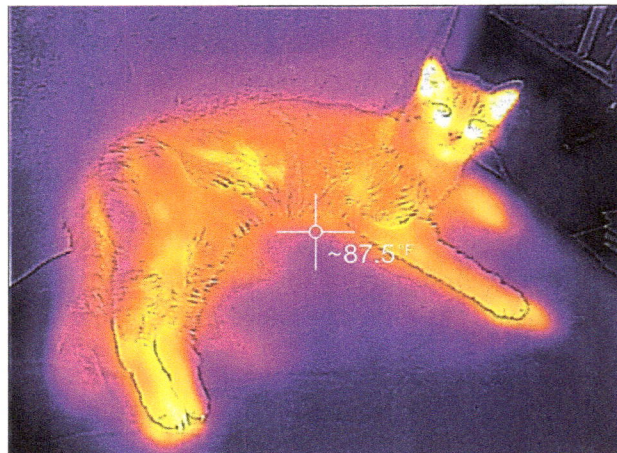

Thermograph of various body parts of a cat.

Cats are familiar and easily kept animals, and their physiology has been particularly well studied; it generally resembles those of other carnivorous mammals, but displays several unusual features probably attributable to cats' descent from desert-dwelling species.

Heat Tolerance

Cats are able to tolerate quite high temperatures. Humans generally start to feel uncomfortable when their skin temperature passes about 38 °C (100 °F), but cats show no discomfort until their skin reaches around 52 °C (126 °F), and can tolerate temperatures of up to 56 °C (133 °F) if they have access to water.

Temperature Regulation

Cats conserve heat by reducing the flow of blood to their skin and lose heat by evaporation through their mouths. Cats have minimal ability to sweat, with glands located primarily in their paw pads, and pant for heat relief only at very high temperatures (but may also pant when stressed). A cat's body temperature does not vary throughout the day; this is part of cats' general lack of circadian rhythms and may reflect their tendency to be active both during the day and at night.

Water Conservation

Cats' feces are comparatively dry and their urine is highly concentrated, both of which are adaptations to allow cats to retain as much water as possible. Their kidneys are so efficient, they can survive on a diet consisting only of meat, with no additional water. They can tolerate high levels of salt only in combination with freshwater to prevent dehydration.

Ability to Swim

While domestic cats are able to swim, they are generally reluctant to enter water as it quickly leads to exhaustion.

Cheetah Anatomy and Physiology

Cheetahs are classified in the family Felidae, subfamily Acinonychinae as the genus *Acynonyx*, and species *jubatus*. The genus contains a single living species.

Cheetahs, one of the bigger free-living cats but at the bottom of the pecking order, are unique animals under threat of extinction because of the increasing loss of their habitat and access to prey species.

They generally appear to have a very limited genetic pool and this genetic uniformity is believed to be the result of at least two population bottlenecks followed by natural inbreeding. The first and most extreme bottleneck possibly occurred in the late Pleistocene (circa 10,000 years ago), while the second was more recent (within the last century) and led to the development of the current South African populations.

Compared to other mammalian species including felids, cheetahs have low levels of MHC (major histocompatibility complex proteins) diversity, but this phenomenon does not seem to influence its immunocompetence and resistance to diseases of free-ranging cheetahs. Additionally, carnivores in general exhibit significantly lower levels of genetic variation than other mammals do, and several carnivore species for which data are available, exhibit lower levels of heterozygosity and polymorphism than cheetahs do. There is speculation though that this limited genetic variation may be responsible for the perceived vulnerability of cheetahs in captivity and in the wild.

There is a belief that cheetahs, as a consequence of their restricted genetic diversity, tend easily to develop clinical signs of inbreeding depression. These are characterised by high neonatal mortality rates that are partially attributed to their perceived increased susceptibility to infections, the difficulty of breeding them in captivity, and the high frequency of spermatozoal defects in captive, and free-ranging cheetah males.

Many of the phenotypic effects that can be attributed to inbreeding depression, such as infertility, reduced litter sizes, and increased susceptibility to disease are, however, normally limited to captive individuals and this phenomenon may be explained as being physiological or behavioural artefacts of captivity, often as a result of inadequate diets and exposure to stress. Usually, however, the survival of cubs in the wild too is poor and there is an expected high attrition rate because of a number of factors, predation being one of the more important ones.

Cheetahs, as a consequence of this perceived inbreeding, also developed metabolic patterns that are substantially different from those of other cats. These variations provide major challenges when attempting to compile balanced diets for cheetahs in captivity as their needs differ substantially from those of the other cats. Nutritional deficiencies and imbalances because of inadequate diets in captivity result in increased neonatal mortalities, poor breeding performance, and various fatal conditions in adult animals.

The current contention is that the genetic constitution of cheetahs does not compromise its survival in the wild and plays a limited role in the poor performance of cheetahs in captivity. Currently the genetic diversity of a metapopulation managed across small reserves is more diverse than that of the free-living population. It is thus important to have genetic information of groups of cheetahs available to allow population management of this threatened species in conservation areas and in small game farms to sustain as much of the genetic variation as is possible.

The extent and geographic patterns of molecular genetic diversity of the largest remaining free-ranging cheetah populations reflect limited differentiation among regions and a generally panmictic population (panmixia or panmixis means random mating). This assumes that there are no mating restrictions, neither genetic nor behavioural, upon the population, and that therefore all recombinations are possible. Where the genetic diversity has been assessed, measures of genetic variation are similar in cheetahs in all regions of Namibia and they are comparable to Eastern African cheetah populations.

In the small and fragmented South African populations, genetic analyses indicate that cheetahs on reserves would not benefit from cross-breeding with the free-roamers but that the free-roamers would benefit. As the reserve population approaches its overall capacity, mating suppression will be required to avoid selling wild cheetah into captivity. Genetic information should continue to be utilized for management of meta-populations, in order to further increase the population's biological fitness over the long term.

Physical Characteristics

Male and Female

Cheetahs are tall and slender animals, with their bodies high off the ground and with long thin legs. Their long tails that may have circular black markings, are distinctive. They have relatively small rounded heads, short muzzles and round ears, and a distinctive black stripe (tear mark) from the inner aspects of the eyes to the corner of their mouths. Males are larger than females with an average weight of 54 kg in males and 43 kg in females. Their total length (including the tail) varies between 2060 cm and 1900 cm in males and females, respectively. Their coat is rough and contains numerous round to oval black spots. The background colour is buffy-white but it varies in intensity according to the region and habitat in which they occur. In desert areas, their colour may be quite pale causing them to blend into the colour of the desert sand.

Cheetah cubs are distinctly different in appearance from the adults. They are less distinctly spotted, and have a rug of long lighter hair on the back, sometimes said, to mimic the coat pattern of honey badgers (*Mellivora capensis*) who are known to be very aggressive animals. This is considered a form of crypsis (crypsis is the use of anatomy and behaviour to hide from potential predators. Cryptic animals are often otherwise palatable to their predators so would never survive if obvious).

The cheetah is an atypical and the most cursorial (having limbs adapted for running) felid. Generally, to maximize its speed, an animal must rapidly swing its limbs (to increase stride frequency) and support its body weight by resisting large ground reaction forces. As a predator, the cheetah also uses its forelimbs for prey capture and they must therefore also be adapted for this function. Its claws are somewhat dog-like both in shape and the diminished degree of retraction that is intermediate between that of other felids and of wolves. They are well known, with the exception of the dew claw (the first digit which is also retractile), for having blunt, only slightly curved, partly retractile claws, considered to be an adaptation for high-speed locomotion.

Some of the morphological differences in the middle phalanges of the cheetah's front feet can be associated with its distinctive hunting behaviour. The reduced manipulative capabilities of the forelimb associated with the evolution of cursorial adaptations seem to have limited the roles of the forepaws in both subduing its prey and feeding. Cheetahs lack the strength of other felids and they are unable to fight their prey to the ground. Instead they must trip or pull them off balance by hooking the rump of its prey and pulling them off balance with their dew claw while running at high speeds.

Cheetahs make various sounds depending on the situation in which they find themselves. When challenged they make a sudden spitting sound reflecting aggression. Normally they communicate with chirping sounds used to locate cubs and members of coalitions. They purr like domestic cats when pleased.

Running Ability and Speed

In nature predator–prey interactions are fundamental in the evolution and structure of ecological communities and within this context cheetahs have evolved the ability to run at speed to capture their prey. Although cheetahs and racing greyhounds are of a similar size and gross morphology, cheetahs are able to achieve far higher top speeds. They are the best sprinters on earth and are capable of accelerating at a rate of up to 7.5 m/s, and to reach speeds in excess of 28 m/s, which is about double that of grey hounds that are considered to be excellent sprinters.

The cheetah is capable of a top speed of about 29 m/s (1749,40 m/min) compared to the maximum speed of 17 m/s achieved by racing greyhounds, and they sprint at speeds that may exceed 100 km/h. However, most cheetah hunts involve only moderate speeds. During a sprint heat production due to muscular activity escalates by about 60 times the rate of heat production at rest. Most of this heat (70%) is stored, the normal evaporative heat loss mechanism not being activated while running. Cheetahs stop running when their rectal body temperature reaches 40.5 C. It thus appears that their body temperature determines the distance that they run after which they abandon the chase, if not successful, after a few hundred metres. Females and trained individuals reach significantly higher speeds compared to males and untrained individuals, respectively.

The cheetah possesses several unique adaptations for high-speed locomotion and fast acceleration, when compared to racing greyhounds. Their hind limb bones are proportionally longer and heavier than in grey hounds, enabling the cheetah to take longer strides. It has a smaller volume of hip extensor musculature than the greyhound, suggesting that the cheetah powers acceleration by using its extensive back musculature. This contention is supported by the presence of extremely powerful psoas muscles that could help to resist the pitching moments around the hip associated with fast acceleration. There is also a proximal to distal reduction in muscle mass on the long bones of the legs, with many of the distal muscles being in series with long tendons. This configuration reduces the inertia of the limb and thus the amount of muscular work required to swing the limb.

It is assumed that these high speeds enable cheetahs to run down slower prey, with failed hunts being attributable to exhaustion and overheating. However, their running pattern during a chase reflects a different hunting strategy. The chase involves alternating between forward and lateral acceleration the extent of which varies according to the prey species. Thus during a chase, they first accelerate to decrease the distance to their prey, before reducing speed 5–8 s from the end of the hunt to enable them to turn rapidly to match prey escape tactics that can involve sudden directional changes and that are difficult to accommodate with an increasing velocity. Moreover, turns at higher speeds lead to greater forces on animals' limbs and muscles, particularly when turn angles are acute, and increase the risk of injury. Thus, while the ability to hunt at high speed may enable cheetahs to outrun their prey, they may not always choose to run at maximum speed, especially when chasing prey species that attempt evasion by sudden changes in direction. These hunting strategies require specific anatomical configurations to allow them to generate speed, and to deal with the torque forces generated by the high speed and sudden changes in direction caused by the typical evasive actions of their prey.

The ability of cheetahs to run very fast is the result of a number of anatomical and physiological parameters including:

Stride Length

Cheetahs have an abnormally long stride that contributes substantially to the speeds achieved when running fast. The configuration of its skeletal and muscle structures increases the large, angular movements of the limb joints that, with bending and straightening of the spine, prolong their stride length. The back muscles, because of their inherent muscle fibre composition, have the ability to produce a strong and quick extension of the spinal column and to increase its rigidity during locomotion thus adding to the force of the forward surge when running.

Stride Frequency

Increasing stride frequency by swinging the limb rapidly and thereby decreasing swing time, also allows quadruped sprinters to reach faster speeds. Cheetahs, however, generally use a lower stride frequency than greyhounds at any given speed and it appears that stride frequency does not contribute substantially to the high speeds reached by cheetahs.

Gait Characteristics

Mammals use two distinct gallops referred to as the transverse (where landing and take-off are contralateral) and rotary (where landing and take-off are ipsilateral). The transverse gallop is characteristically that of the horse, while cheetahs use the rotary gallop. The fundamental difference between these gaits is determined by which set of limbs, fore or hind, initiates the transition of the centre of mass from a downward–forward to upward–forward trajectory that occurs between the main portions of the stride when the animal makes contact with the ground.

During a stride when their feet are in contact with the ground, animals support their body weight by resisting joint torques caused by the impact. Quadrupeds typically support a greater proportion of their body weight with their forelimbs during steady state locomotion. At high speed, however, such as in cheetahs running at speed, the hind limbs support the majority of an animal's body weight, and within this context, cheetahs support 70% of their body weight on their hind limbs at speeds of 18 m/s. The hindquarters thus generate most of the muscular activity required for propulsion. It is assumed that supporting a greater proportion of body weight on a particular limb is also likely to reduce the risk of slipping during propulsive efforts.

When travelling at top speed the cheetah's forelimbs also experience very high peak forces, and they too must be able to cope with large joint torques but they do not contribute substantially to generating the superior speed of cheetahs.

Muscle Structure and Conformation

The distribution of skeletal muscle along the legs plays a role in animals that have the ability to reach high speeds and have to accommodating high impact pressures during a chase. In cheetahs the proximal limbs contain many, large, PCSA (physiological cross-sectional area) muscles. This configuration provides the limbs with the ability to resist and absorb large impacts. The legs, because of these muscle masses that can absorb some of the impact, do not merely function as simple struts and, because of the muscle distribution, they can absorb much of the force of impact while running. This conformation also provides cheetahs with the ability to control and stabilise their legs during high-speed manoeuvring such as is needed during a chase. The large digital flexors and extensor muscles characteristic of the cheetah's forelimb, allow it to dig its digits into the ground, providing the required traction when galloping and manoeuvring at speed.

Muscle Fibre Composition and Characteristics

Muscle consists of muscle fibres (Type I and Type II fibres) that have different functional and metabolic characteristics. The characteristic species-specific variations in fibre composition reflect the physiological needs of individual animal species, and in felids generally the fibre type composition of hind limb muscles matches their daily activity patterns. Roaming tigers, for instance, walk long distances, while cheetahs have requirements for speed and power over short distances. There is a relationship between the amount and type of activity, and the myosin heavy chain (MHC) isoform composition of a muscle in that tigers have a high combined percentage of the characteristically slower-twitch fibre isoforms required for sustained activity (MHCs I and IIa).

Cheetah locomotory muscles contain a high proportion of fast-twitch fibres, needed for rapidly swinging their limbs and reducing limb swing-times required for running at speed over short distances. In their hind limb muscles there is a higher percentage of Type II (Type IIa + IIx) fibres than in the forelimb muscles further confirming that the propulsive role of the hind limb is greater than the forelimb.

Enzyme activity in cheetah muscle reflects a high capacity for glycolysis in anaerobically based exercise required for sprinting during high-speed chases for hunting purposes. The fibre type composition, mitochondrial content, and glycolytic enzyme capacities in the locomotory muscles of cheetahs operate at the extreme range of values for other sprinters bred or trained for this activity including greyhounds, thoroughbred horses and human athletes.

King Cheetahs

The iconic and spectacular king cheetah has become a well-known feature of the cheetahs at HESC.

The first recorded description of the king cheetah was from the Macheke district in Zimbabwe in 1927 where it was thought to be a hybridisation of a leopard and a cheetah. There the local indigenous population, who had known of its existence for a long time, referred to it as nsuifisi, the hyena-leopard, and told many tales of its fierceness. At a time, they seemed to be more common in parts of Zimbabwe where the colonists in the then Rhodesia referred to them as Mazoe leopards.

THE KING CHEETAH

Now it is known that what was once considered a separate species (*Acinonyx rex*), is but one of many colour variants that have been described in cheetahs. Other such variants include albinistic, melanistic, cream (isabelline), black with ghost markings, and red (erythristic) with dark tawny spots on a golden background. Blue (or grey) cheetahs have been described as white cheetahs with grey-blue spots (chinchilla) or pale grey cheetahs with darker grey spots (Maltese mutation). Some desert region cheetahs are unusually pale probably because it provides a better camouflage in the like-coloured desert.

King cheetahs are infrequently seen in the wild. The last recorded sighting of a king cheetah in the wild was in 1986 in the Kruger National Park. They occur naturally in a localised area that covers adjoining portions of Botswana, Zimbabwe, and South Africa (northern and eastern regions of the Limpopo Province).

During the 1980s, a number of litters born in captivity contained king cheetah cubs, and since then it has become customary for some of the breeding facilities (De Wildt and HESC) to focus on sustaining blood lines with the intention of breeding king cheetahs at will. The gene appears to be carried at a low frequency in the wild, and its occurrence is localised.

The king coat colour pattern is the consequence of a mutation at the tabby locus inherited as a single, autosomal, recessive allele.

Adaptions of the Cheetah

A cheetah chase is not all that eager to be dinner, so when it is in a race with the worlds fastest land animal it just makes sense not to run in a straight line. Football players know this; they change direction often so they aren't tackled easily. But, the cheetah is ready for this tactic. Claws that stick out like a dog's give the cheetah traction in high speed turns. Even the cheetah's tail helps. Other cats have round, fluffy tails like your house cat but not the cheetah. Its tail has a flat surface, like the rudder of a boat, and it helps balance its body as the cheetah runs.

So the cheetah has speed licked. But there is another problem stopping. When the antelope falls to the ground, tackled by the cheetah, the cat is still going 60 mph. The antelope isn't going to wait around if the cheetah flies by it. In order to stop immediately, the cheetah has a highly specialized, pointed pad in the back of each front leg. So, while going full speed, the cheetah can slam its two front legs down, hard. The pads tear into the ground bringing the cat to a near instantaneous halt. Then, it grabs its dinner before the antelope can get away. What a game of predator and prey. Both animals are equipped to survive. That is what the balance of life is all about.

Cheetah's, like most animals in a competitive environment, have other adaptations developed overtime to create their niche and ensure their survival.

Cheetah's great speed comes from their long stride and limber but muscular legs.This differentiates them from the other Cat families such as the lions and tigers.Other competitors include hyenas and wild dogs.The agility and high speed of a cheetah also allows them to attack alone, unlike the hyenas, lions and dogs who attack in packs. Their prey also differs in that they mostly are more slender and quicker animals. The gazelle, springbok, impala and antelope are common prey for the cheetah.

This means that they are also easier to chew and so only need smaller teeth but in turn cheetahs have a bigger nasal cavities that allow it to suck greater air needed as they sprint for their prey. Cheetahs have a high success rate when chasing prey but they do lose a significant amount of their catch to other larger animals such as the lion.It is not common for them to defend their kill.

Cheetahs other adaption is their color .Their variegated dark spots on a light background help camouflage them to some degree.

Have you noticed that footballers and other sportsman will have dark stripes under their eyes to prevent glare from either the Sun or bright lights. Well cheetahs have this too. Those black lines down their face are called "tear marks" and they are also to prevent glare from the hot overhead Sun.

Reproduction

In order to understand how cheetahs coexist and compete with other species in their environment, it's important to know a few things about their behavior. Given the rapid decline in their numbers over the last two decades, it might be surprising to some that the cheetah is the most reproductive of all cats. Male cheetahs are not involved with caring for their young. In fact, the only time a female cheetah has interaction with a male is for the purposes of mating only. After mating, the gestation period and caring for the cubs are the sole responsibility of the female.

A female is normally only pregnant for approximately three months before giving birth and litters are typically three to five cubs. They can sometimes be larger in number but the biggest cheetah litter that has been documented was eight cubs. While the short gestation period and relatively large litters seem like they would provide a fair amount of support to growing the cheetah population, other factors have significant negative consequences that make things difficult for cubs to survive.

Because the female cheetah handles the raising of the cubs alone, it means she must leave the cubs alone quite often while she hunts for food. This leaves young cubs virtually defenseless against other predators that come along such as lions, hyenas and even large predatory birds. According to the Wildlife Conservation Society, only 10% of all the cheetahs born make it past three months of age. Fifty percent of cubs are killed by other large predators and forty percent die from a wide range of diseases due to a weak immune system.

If food and resources are too difficult to find, a female cheetah will many times leave her cubs and abandon them so she can survive. If that does not happen, then cubs are typically weaned by the time they are six to eight weeks old and they are ready to leave the mother once they are fifteen months to two years old.

With the high mortality of cubs and the passive-aggressive nature of cheetahs, it makes it difficult for them to repopulate and directly compete with larger predators.

References

- Mintz, Zoe (14 January 2014). "Humans And Primates Burn 50 Percent Fewer Calories Each Day Than Other Mammals". Www.ibtimes.com. IBT Media Inc. Retrieved 2014-01-14

- Form-and-function, mammal, animal: britannica.com, Retrieved 4 July, 2019

- Horvath, J.; et al. (2008). "Development and Application of a Phylogenomic Toolkit: Resolving the Evolutionary History of Madagascar's Lemurs". Genome Research. 18 (3): 489–499. Doi:10.1101/gr.7265208. PMC 2259113. PMID 18245770

- H Groves, C.P. (2005). Wilson, D.E.; Reeder, D.M. (eds.). Mammal Species of the World: A Taxonomic and Geographic Reference (3rd ed.). Baltimore: Johns Hopkins University Press. Pp. 111–184. ISBN 0-801-88221-4. OCLC 62265494

- Gouteux, S.; Thinus-Blanc, C.; Vauclair, J. (2001). "Rhesus monkeys use geometric and nongeometric information during a reorientation task" (PDF). Journal of Experimental Psychology: General. 130 (3): 505–519. Doi:10.1037/0096-3445.130.3.505

- Bouchet H, Pellier A, Blois-Heulin C, Lemasson A. 2010. Sex differences in the vocal repertoire of adult red-capped mangabeys (Cercocebus torquatus): A multi-level acoustic analysis. American Journal of Primatology 72:360-375

- Lindenfors, P. (2002). "Sexually antagonistic selection on primate size". Journal of Evolutionary Biology. 15 (4): 595–607. Doi:10.1046/j.1420-9101.2002.00422.x. ISSN 1010-061X

- Cheetah: lions.org, Retrieved 5 August, 2019

The Behavior of Mammals — 4

Behavior is the way in which a living organism acts or conducts itself due to the stimulus presented by its surrounding. Behavioral aspects of mammals such as cetacea, equine, felidae, etc. are thoroughly discussed in this chapter. This chapter has been carefully written to provide an easy understanding of the various concepts related to behavior of mammals.

Mammals Behavior

Social Behavior

The dependence of the young mammal on its mother for nourishment has made possible a period of training. Such training permits the nongenetic transfer of information between generations. The ability of young mammals to learn from the experience of their elders has allowed a behavioral plasticity unknown in any other group of organisms and has been a primary reason for the evolutionary success of mammals. The possibility of training is one of the factors that has made increased brain complexity a selective advantage. Increased associational potential and memory extend the possibility of learning from experience, and the individual can make adaptive behavioral responses to environmental change. Individual response to short-term change is far more efficient than genetic response.

Some types of mammals are solitary except for brief periods when the female is in estrus. Others, however, form social groups. Such groups may be reproductive or defensive, or they may serve both functions. Within the social group, the hierarchy may be maintained through physical combat between individuals, but in many cases stereotyped patterns of behaviour evolve to displace actual combat, thereby conserving energy while maintaining the social structure.

Leopard (Panthera pardus) grooming her cub in
the Masai Mara National Reserve, Kenya.

A pronounced difference between sexes (sexual dimorphism) is frequently extreme in social mammals. In large part this is because dominant males tend to be those that are largest or best-armed. Dominant males also tend to have priority in mating or may even have exclusive responsibility for mating within a "harem." Rapid evolution of secondary sexual characteristics, including size, can take place in a species with such a social structure.

A complex behaviour termed "play" frequently occurs between siblings, between members of an age class, or between parent and offspring. Play extends the period of maternal training and is especially important in social species, providing an opportunity to learn behaviour appropriate to the maintenance of dominance.

Territoriality

That area covered by an individual in its general activity is frequently termed the home range. A territory is a part of the home range defended against other members of the same species. As a generalization it may be said that territoriality is more important in the behaviour of birds than of mammals, but data for the latter are available primarily for diurnal species. Frequently territories of mammals are marked, either with urine or with secretions of specialized glands, as in lemurs. This form of territorial labeling is less evident to humans than the singing or visual displays of birds. Many mammals that do not maintain territories per se nevertheless will not permit unlimited crowding and will fight to maintain individual distance. Such mechanisms result in more economical spacing of individuals over the available habitat.

Yellowstone wolf pack.

Primate Behavior

Social Systems

Richard Wrangham stated that social systems of non-human primates are best classified by the amount of movement by females occurring between groups. He proposed four categories:

- Female transfer systems – Females move away from the group in which they were born. Females of a group will not be closely related whereas males will have remained with their natal groups, and this close association may be influential in social behavior. The groups formed are generally quite small. This organization can be seen in chimpanzees, where the males, who are typically related, will cooperate in defense of the group's territory. Among New World Monkeys, spider monkeys and muriquis use this system.

A social huddle of ring-tailed lemurs. The two individuals on the right exposing their white ventral surface are sunning themselves.

- Male transfer systems – While the females remain in their natal groups, the males will emigrate as adolescents. Polygynous and multi-male societies are classed in this category. Group sizes are usually larger. This system is common among the ring-tailed lemur, capuchin monkeys and cercopithecine monkeys.

- Monogamous species – A male–female bond, sometimes accompanied by a juvenile offspring. There is shared responsibility of parental care and territorial defense. The offspring leaves the parents' territory during adolescence. Gibbons essentially use this system, although "monogamy" in this context does not necessarily mean absolute sexual fidelity. These species do not live in larger groups.

- Solitary species – Often males who defend territories that include the home ranges of several females. This type of organization is found in the prosimians such as the slow loris. Orangutans do not defend their territory but effectively have this organization.

Other systems are known to occur as well. For example, with howler monkeys both the males and females typically transfer from their natal group on reaching sexual maturity, resulting in groups in which neither the males nor females are typically related. Some prosimians, colobine monkeys and callitrichid monkeys use this system.

The transfer of females or males from their native group is likely an adaptation for avoiding inbreeding. An analysis of breeding records of captive primate colonies representing numerous different species indicates that the infant mortality of inbred young is generally higher than that of non-inbred young. This effect of inbreeding on infant mortality is probably largely a result of increased expression of deleterious recessive alleles.

The chimpanzees are social great apes.

Primatologist Jane Goodall, who studied in the Gombe Stream National Park, noted fission-fusion societies in chimpanzees. There is fission when the main group splits up to forage during the day, then fusion when the group returns at night to sleep as a group. This social structure can also be observed in the hamadryas baboon, spider monkeys and the bonobo. The gelada has a similar social structure in which many smaller groups come together to form temporary herds of up to 600 monkeys.

These social systems are affected by three main ecological factors: distribution of resources, group size, and predation. Within a social group there is a balance between cooperation and competition. Cooperative behaviors include social grooming (removing skin parasites and cleaning wounds), food sharing, and collective defense against predators or of a territory. Aggressive behaviors often signal competition for food, sleeping sites or mates. Aggression is also used in establishing dominance hierarchies.

Interspecific Associations

Several species of primates are known to associate in the wild. Some of these associations have been extensively studied. In the Tai Forest of Africa several species coordinate anti-predator behavior. These include the Diana monkey, Campbell's mona monkey, lesser spot-nosed monkey, western red colobus, king colobus (western black and white colobus), and sooty mangabey, which coordinate anti-predator alarm calls. Among the predators of these monkeys is the common chimpanzee.

The red-tailed monkey associates with several species, including the western red colobus, blue monkey, Wolf's mona monkey, mantled guereza, black crested mangabey and Allen's swamp monkey. Several of these species are preyed upon by the common chimpanzee.

In South America, squirrel monkeys associate with capuchin monkeys. This may have more to do with foraging benefits to the squirrel monkeys than anti-predation benefits.

Cognition and Communication

Pair of black howler monkeys vocalizing.

Primates have advanced cognitive abilities: Some make tools and use them to acquire food and for social displays; some can perform tasks requiring cooperation, influence and rank; they are status conscious, manipulative and capable of deception; they can recognise kin and conspecifics; and they can learn to use symbols and understand aspects of human language including some relational syntax and concepts of number and numerical sequence. Research in primate cognition explores problem solving, memory, social interaction, a theory of mind, and numerical, spatial, and abstract concepts. Comparative studies show a trend towards higher intelligence going from prosimians to New World monkeys to Old World monkeys, and significantly higher average cognitive abilities in the great apes. However, there is a great deal of variation in each group (e.g., among New World monkeys, both spider and capuchin monkeys have scored highly by some measures), as well as in the results of different studies.

Lemurs, lorises, tarsiers, and New World monkeys rely on olfactory signals for many aspects of social and reproductive behavior. Specialized glands are used to mark territories with pheromones, which are detected by the vomeronasal organ; this process forms a large part of the communication behavior of these primates. In Old World monkeys and apes this ability is mostly vestigial, having regressed as trichromatic eyes evolved to become the main sensory organ. Primates also use vocalizations, gestures, and facial expressions to convey psychological state. Facial musculature is complex in primates, particularly in monkeys and apes. Like humans, chimpanzees can distinguish the faces of familiar and unfamiliar individuals. Hand and arm gestures are also important forms of communication for great apes and a single gesture can have multiple functions.

The Philippine tarsier, has a high-frequency limit of auditory sensitivity of approximately 91 kHz with a dominant frequency of 70 kHz. Such values are among the highest recorded for any terrestrial mammal, and a relatively extreme example of ultrasonic communication. For Philippine tarsiers, ultrasonic vocalizations might represent a private channel of communication that subverts detection by predators, prey and competitors, enhances energetic efficiency, or improves detection against low-frequency background noise. Male howler monkeys are among the loudest land mammals and their roars can be heard up to 4.8 km (3.0 mi). Roars are produced by modified larynx and enlarged hyoid bone which contains an air sac. These calls are thought to relate to intergroup spacing and territorial protection as well as possibly mate-guarding. The vervet monkey gives a distinct alarm call for each of at least four different predators, and the reactions of other monkeys vary according to the call. For example, if an alarm call signals a python, the monkeys climb into the trees, whereas the eagle alarm causes monkeys to seek a hiding place on the ground. Many non-human primates have the vocal anatomy to produce human speech but lack the proper brain wiring. Vowel-like vocal patterns have been recorded in baboons which has implications for the origin of speech in humans.

Aggression

Non-human primates and humans have been observed to be very similar in terms of personality, such as chimpanzees having "'Big Five' personality factors found in humans, i.e. neuroticism, extraversion, openness, agreeableness, and conscientiousness". Primates seem to possess a sixth personality trait, dominance. Chimpanzees and bonobos tend to be more social than male orangutans, who cannot be in the same living space. Male gorillas display aggression during a majority of their interactions while chimpanzees reconcile after their aggressive encounters. Killing males in differing groups in order to expand territory is an act of violence only seen in chimps and humans. Bonobos tend to live in cohesive communities with lower aggressive incidents. This reduction in reactive aggression committed by male bonobos may be due to the evolution of prominent female groups.

Both humans and nonhuman primates, such as chimpanzees, exhibit proactive aggression, a type of preplanned aggression with a reward. This aggression is expressed between neighboring groups. Proactive aggression ultimately increases the fitness of the community as a whole as the size of the community increases. Unlike proactive aggression, reactive aggression is low in humans but high in chimpanzees. Reactive aggression is a result of anger in order to cease a stressful stimulus. Low reactive aggression in humans can be attributed to tolerance and cooperation. Wrangham found "evolution of within-group tolerance, such as individual selection for cooperative breeding, group selection for parochial altruism, and cultural group selection for prosocial norms". Ranking in nonhuman primates stems from the most aggressive male, while nomadic hunter-gatherers are respected for their prestige and ability to form alliances and negotiations. Boehm proposed the "execution hypothesis", or the approval by the community of killing group members who violate the norm of the group or are deemed a bully. Hunter-gatherers who attempted to utilize aggression were viewed as a threat and removed from the group. The removal of community threats can be likened to modern capital punishment, more specifically the death penalty.

Life History

Primates have slower rates of development than other mammals. All primate infants are breastfed by their mothers (with the exception of some human cultures and various zoo raised primates which are fed formula) and rely on them for grooming and transportation. In some species, infants are protected and transported by males in the group, particularly males who may be their fathers. Other relatives of the infant, such as siblings and aunts, may participate in its care as well. Most primate mothers cease ovulation while breastfeeding an infant; once the infant is weaned the mother can reproduce again. This often leads to weaning conflict with infants who attempt to continue breastfeeding.

Infanticide is common in polygynous species such as gray langurs and gorillas. Adult males may kill dependent offspring that are not theirs so the female will return to estrus and thus they can sire offspring of their own. Social monogamy in some species may have evolved to combat this behavior. Promiscuity may also lessen the risk of infanticide since paternity becomes uncertain.

A crab-eating macaque breastfeeding her baby.

Primates have a longer juvenile period between weaning and sexual maturity than other mammals of similar size. Some primates such as galagos and new world monkeys use tree-holes for nesting, and park juveniles in leafy patches while foraging. Other primates follow a strategy of riding, i.e. carrying individuals on the body while feeding. Adults may construct or use nesting sites, sometimes accompanied by juveniles, for the purpose of resting, a behavior which has developed secondarily in the great apes. During the juvenile period, primates are more susceptible than adults to predation and starvation; they gain experience in feeding and avoiding predators during this time. They learn social and fighting skills, often through playing. Primates, especially females, have longer lifespans than other similarly sized mammals, this may be partially due to their slower metabolisms. Late in life, female catarrhine primates appear to undergo a cessation of reproductive function known as menopause; other groups are less studied.

Diet, Feeding and Hunting

Leaf eating mantled guereza, a species of black-and-white colobus.

A mouse lemur holds a cut piece of fruit in its hands and eats.

Primates exploit a variety of food sources. It has been said that many characteristics of modern primates, including humans, derive from an early ancestor's practice of taking most of its food from the tropical canopy. Most primates include fruit in their diets to obtain easily digested nutrients including carbohydrates and lipids for energy. Primates in the suborder Strepsirrhini (non-tarsier prosimians) are able to synthesize vitamin C, like most other mammals, while primates of the suborder Haplorrhini (tarsiers, monkeys and apes) have lost this ability, and require the vitamin in their diet.

Many primates have anatomical specializations that enable them to exploit particular foods, such as fruit, leaves, gum or insects. For example, leaf eaters such as howler monkeys, black-and-white colobuses and sportive lemurs have extended digestive tracts which enable them to absorb nutrients from leaves that can be difficult to digest. Marmosets, which are gum eaters, have strong incisor teeth, enabling them to open tree bark to get to the gum, and claws rather than nails, enabling them to cling to trees while feeding. The aye-aye combines rodent-like teeth with a long, thin middle finger to fill the same ecological niche as a woodpecker. It taps on trees to find insect larvae, then gnaws holes in the wood and inserts its elongated middle finger to pull the larvae out. Some species have additional specializations. For example, the grey-cheeked mangabey has thick enamel on its teeth, enabling it to open hard fruits and seeds that other monkeys cannot. The gelada is the only primate species that feeds primarily on grass.

Hunting

Humans have traditionally hunted prey for subsistence.

Tarsiers are the only extant obligate carnivorous primates, exclusively eating insects, crustaceans, small vertebrates and snakes (including venomous species). Capuchin monkeys can exploit many different types of plant matter, including fruit, leaves, flowers, buds, nectar and seeds, but also eat insects and other invertebrates, bird eggs, and small vertebrates such as birds, lizards, squirrels and bats.

The common chimpanzee eats an omnivorous frugivorous diet. It prefers fruit above all other food items and even seeks out and eats them when they are not abundant. It also eats leaves and leaf buds, seeds, blossoms, stems, pith, bark and resin. Insects and meat make up a small proportion of their diet, estimated as 2%. The meat consumption includes predation on other primate species, such as the western red colobus monkey. This sometimes involves tool use. Common chimpanzees sharpen sticks to use as weapons when hunting mammals. This is considered the first evidence of systematic use of weapons in a species other than humans. Researchers documented 22 occasions where wild chimpanzees fashioned sticks into "spears" to hunt lesser bush babies (*Galago senegalensis*). In each case, a chimpanzee modified a branch by breaking off one or both ends and, frequently using its teeth, sharpened the stick. The tools, on average, were about 60 cm (24 in) long and 1.1 cm (0.4 in) in circumference. The chimpanzees then jabbed their spears into hollows in tree trunks where bush babies slept. There was a single case in which a chimpanzee successfully extracted a bush baby with the tool. The bonobo is an omnivorous frugivore – the majority of its diet is fruit, but it supplements this with leaves, meat from small vertebrates, such as anomalures, flying squirrels and duikers, and invertebrates. In some instances, bonobos have been shown to consume lower-order primates.

As Prey

Predators of primates include various species of carnivorans, birds of prey, reptiles, and other primates. Even gorillas have been recorded as prey. Predators of primates have diverse hunting strategies and as such, primates have evolved several different antipredator adaptations including crypsis, alarm calls and mobbing. Several species have separate alarm calls for different predators such as air-borne or ground-dwelling predators. Predation may have shaped group size in primates as species exposed to higher predation pressures appear to live in larger groups.

Tool use and Manufacture

Tool Use

A western lowland gorilla using a stick possibly to gauge the depth of water.

Crab-eating macaques with stone tools.

There are many reports of non-human primates using tools, both in the wild or when captive. The use of tools by primates is varied and includes hunting (mammals, invertebrates, fish), collecting honey, processing food (nuts, fruits, vegetables and seeds), collecting water, weapons and shelter.

In 1960, Jane Goodall observed a chimpanzee poking pieces of grass into a termite mound and then raising the grass to his mouth. After he left, Goodall approached the mound and repeated the behaviour because she was unsure what the chimpanzee was doing. She found that the termites bit onto the grass with their jaws. The chimpanzee had been using the grass as a tool to "fish" or "dip" for termites. There are more limited reports of the closely related bonobo using tools in the wild; it has been claimed they rarely use tools in the wild although they use tools as readily as chimpanzees when in captivity. It has been reported that females, both chimpanzee and bonobo, use tools more avidly than males. Orangutans in Borneo scoop catfish out of small ponds. Anthropologist Anne Russon saw several animals on these forested islands learn on their own to jab at catfish with sticks, so that the panicked prey would flop out of ponds and into the orangutan's waiting hands There are few reports of gorillas using tools in the wild. An adult female western lowland gorilla used a branch as a walking stick apparently to test water depth and to aid her in crossing a pool of water. Another adult female used a detached trunk from a small shrub as a stabilizer during food gathering, and another used a log as a bridge.

The black-striped capuchin was the first non-ape primate for which tool use was documented in the wild; individuals were observed cracking nuts by placing them on a stone anvil and hitting them with another large stone. In Thailand and Myanmar, crab-eating macaques use stone tools to open nuts, oysters and other bivalves, and various types of sea snails. Chacma baboons use stones as weapons; stoning by these baboons is done from the rocky walls of the canyon where they sleep and retreat to when they are threatened. Stones are lifted with one hand and dropped over the side whereupon they tumble down the side of the cliff or fall directly to the canyon floor.

Although they have not been observed to use tools in the wild, lemurs in controlled settings have been shown to be capable of understanding the functional properties of the objects they had been trained to use as tools, performing as well as tool-using haplorhines.

Tool Manufacture

Tool manufacture is much rarer than simple tool use and probably represents higher cognitive functioning. Soon after her initial discovery of tool use, Goodall observed other chimpanzees picking up leafy twigs, stripping off the leaves and using the stems to fish for insects. This change of a leafy twig into a tool was a major discovery. Prior to this, scientists thought that only humans manufactured and used tools, and that this ability was what separated humans from other animals. Both bonobos and chimpanzees have also been observed making "sponges" out of leaves and moss that suck up water and are used as grooming tools. Sumatran orangutans have been observed making and using tools. They will break off a tree branch that is about 30 cm long, snap off the twigs, fray one end and then use the stick to dig in tree holes for termites. In the wild, mandrills have been observed to clean their ears with modified tools. Scientists filmed a large male mandrill at Chester Zoo (UK) stripping down a twig, apparently to make it narrower, and then using the modified stick to scrape dirt from underneath its toenails. Captive gorillas have made a variety of tools.

Nest-building in Primates

Nest-building in primates refers to the behaviour of building nests by extant strepsirrhines (lemurs and lorisoids) and hominid apes (chimpanzees, gorillas, orangutans and humans). Strepsirrhines build nests for both sleeping and also for raising families. Hominid apes build nests for sleeping at night, and in some species, for sleeping during the day. Nest-building by hominid apes is learned by infants watching the mother and others in the group, and is considered tool use rather than animal architecture. Old World monkeys and New World monkeys do not nest.

It has been speculated that a major evolutionary advance in the cognitive abilities of hominoids may first have occurred due to the development of nest-building behaviour and that the transition from nest-building to ground-sleeping led to "modifications in the quality and quantity of hominid sleep, which in turn may have enhanced waking survival skills through priming, promoted creativity and innovation, and aided the consolidation of procedural memories".

In Strepsirrhines

Strepsirrhines may be nocturnal, diurnal, or crepuscular. Some may either occupy holes in trees or build nests. Unlike the hominid apes, strepsirrhines build nests by instinct and use them for breeding purposes. Strepsirrhines mothers either carry their young on their bodies, conceal their young in foliage while they venture out to feed, returning periodically to feed and groom them, or leave them in a nest built for that purpose.

In Lemurs

Among lemurs, females of smaller species, such as the mouse lemur and giant mouse lemur, build leaf nests before birth for the protection of young. Leaf nests in golden-brown mouse lemurs may provide thermoregulation benefits. Male mouse lemurs have been found sharing nests with up to seven females at a time during the mating season.

In the Ruffed lemur (*Varecia variegata*), nests are made from locally collected material and may also be lined with hair plucked from their own body. In larger lemur species such as the Verreaux's sifaka, ring-tailed lemur and common brown lemur, the young cling to the mother and nests are not built.

Aye-ayes, being nocturnal, build oval-shaped nests from nearby branches and lianas for day time use. These may be built in a tangle of lianas or in a fork of a tree, at a height of 7 to 20 metres (23 to 66 ft) above the ground. These nests may be re-used by other aye-ayes once the original occupant moves on. A single occupant uses a nest for a few days, refreshing it regularly with fresh vegetation. Aye-ayes build and use many nests during one study, 8 aye-ayes were recorded to build and use as many as a hundred different nests over a period of 2 years.

Some strepsirrhines, such as mouse lemurs, build nests.

In Lorisoids

Lesser bushbaby mothers initially shelter their offspring, usually twins, in a nest or tree hollow, later concealing the infants in foliage while they forage at night. In some species, such as dwarf galagos, the day-sleeping nests may be shared by groups of females or occasionally by visiting males.

In Hominid Apes

Gorilla night nest.

Hominid apes construct nests during the day or by night, primarily for resting. The nests are not built using instinct but through behavioural patterns which are learned by the young from their mother. Nest building is habitual behaviour, and nest-counts and faecal analysis at each nest site can be used to estimate hominid ape population counts and composition. In the case of orangutans and chimpanzees, social influences are probably essential for the animals to develop successful nesting-behaviour.

In Gorillas

Gorillas construct nests for day and night use. Day nests tend to be simple aggregations of branches and leaves on the ground, while night nests are more elaborate constructions usually on the ground but sometimes on trees, especially those of juveniles and females in areas submitted to a high poaching pressure. The nests may be 1 to 5 feet (0.30 to 1.52 m) in diameter and are constructed by individuals. Young animals nest with their mother but do not construct nests until three years of age, initially in close proximity to their mother. Gorilla nests are distributed randomly and the tree species used appears to be opportunistic.

In Chimpanzees and Bonobos

Chimpanzee nest.

Nest-building is seen in chimpanzees and bonobos who construct arboreal night nests by lacing together branches from one or more trees. They also can build nap nests to rest in the afternoon, these are usually more poorly constructed than the night nests and can be built both on the ground and on the trees. At some research sites such as Bili forest, in Congo, chimpanzees can build a significant proportion of their night nests on the ground. Nests consist of a mattress, supported on a strong foundation, and lined above with soft leaves and twigs. Nests are built in trees which have a minimum diameter of 5 centimetres (0.16 ft) and may be located at a height of 1 to 45 metres (3.3 to 147.6 ft). Day and night nests are built. Nests may be located in groups.

In Orangutans

Orangutans build day and night nests. Young orangutans learn by observing their mother's nest-building behaviour. Nest-building is a leading reason for young orangutans leaving their mother for the first time. Starting at 6 months of age, orangutans practice nest building and gain proficiency by the time they are 3 years old. Initially, a suitable tree is located. Orangutans are selective about sites even though many tree species are utilised. The foundation is then built by pulling together branches under them and joining them at a point. After the foundation has been built, they bend smaller, leafy branches onto the foundation; this serves the purpose of and is termed as the "mattress". After this, orangutans braid the tips of branches into the mattress. This increases the stability of the nest and forms the final process of nest building. Orangutans may add additional features such as "pillows", "blankets", "roofs" and "bunk-beds" to their nest. Orangutans make "pillows" by clumping together leafy branches with the leaves in the center and the twig shoots pointed outward. They bite the twigs to blunt sharp ends. Pillows are added to night nests but are usually absent from day nests. A "blanket" consists of large leafy branches with which orangutans cover themselves after lying down. Orangutans may create a waterproof overhead shelter for the nest by braiding together a loose selection of branches. They may also make a "bunk-nest" or "bunk-bed", a few metres above the main nest.

Orangutan nest.

Primate Sociality

Stump-tailed macaques.

Most primates live in groups. The best explanation for why animals form groups and endure the costs of feeding competition is to minimize the risk of predation. Grouping patterns are tied to diet and the defensibility of resources. Females are out to maximize resources for themselves and their offspring, so as to maximize their reproductive success. If a species eats grass or leaves, it does not make sense to defend those resources. However, there is safety in numbers and those species (especially arboreal species) will normally be found living or foraging in small groups. If a species specializes on ripe fruit, they cannot defend them because of the patchy nature of fruit in geographic space and time. In the case of the few primate ripe fruit specialists, such as chimps and spider monkeys, males defend a home range that contains resources that females need, and thus females are attracted to join them. While orangutans are also preferentially frugivorous, they are solitary due to their large size and strict arboreality, which limits resources to those that are accessible from supporting branches. Finally, if a species can eat a variety of things that come in variable-sized patches, they can band together and defend those resources as they come across them in their daily ranging. In that case, females stay together in their natal group (termed female philopatry) and cooperate in resource defense.

Social organization involves several aspects of group life, such as (1) the average numbers of individuals in terms of age and sex; (2) whether group members remain in their natal group at maturity or leave, and hence whether individuals have relatives in the group; (3) whether those animals that join a group in adulthood stay permanently or tend to leave after a period of time; (4) the pattern of interactions between individuals, e.g. whether there is a dominance hierarchy and if so, if an individual's position in the hierarchy is permanent or temporary; and (5) the number of potential mates to which an individual has access. While we tend to categorize species by their grouping pattern or social organization, it is increasingly apparent that there is variability within primate species. Some species share our pattern of living in multi-male/multi-female groups. Other categories of primate social organization are solitary, male-female pairs, and one-male/ multi-female groups. Interestingly, all of the mating systems seen in primates, i.e. monogamy, polygyny (one male mates with multiple females), polyandry (one female mates with multiple males), and polygynandry (both males and females are promiscuous), are also seen in humans. Some men and women marry or mate for life; some men have multiple wives or partners, and the same goes for some women.

Solitary and Dispersed Polygyny

Except for the orangutans, solitary foragers are small nocturnal prosimians that forage primarily for insects and fruit. Examples of solitary foragers are the bushbabies and pottos of Africa, most of the nocturnal lemurs of Madagascar, and the lorises of Asia. Prosimian solitary foragers either avoid predation by stealth (i.e. the slow climbers, such as pottos and slow lorises) or a form of locomotion termed vertical clinging and leaping (e.g., bushbabies) that allows for quick getaways. Females usually forage alone and either park their young nearby or leave them in a "nest," such as a tree hole. Sleeping groups may consist of female relatives and their young and females, young, and males, depending on the species and female-female tolerance. Male home ranges often overlap multiple female home ranges, and males monitor female sexual cycles by "making the rounds" and monitoring their scent, hence the use of the term "dispersed polygyny," i.e. one male and multiple dispersed females. One male may dominate other smaller or less dominant males in an area and may suppress them from breeding, via pheromonal activity.

Bushbaby.

Orangutans are the odd man out. They are large and arboreal so they do not need to group for protection. They need a lot of resources to support them and at some sites, they suffer periodic food shortages, so that grouping would hinder foraging. Females and their dependent offspring forage together. Females maintain proximity and mate with a dominant male with developed secondary sexual characteristics, i.e. large size, a throat sac for loud calls, and facial flanges. Until there is an opportunity for males to acquire females, such as when a large male dies, males stay small and mate opportunistically. Scientists are stymied at how they can delay maturation and then facultatively develop into the larger morph.

Territorial Pairs and Monogamy

While a few species of primates are commonly referred to as monogamous, extrapair copulations have been observed in every one of them. The last primate to have lost the title of true monogamist was the night monkey of Central and South America.

Gibbon of Southeast Asia. "Gibbon Hoolock de l'ouest".

Monogamy begs the question, "why?" While females may benefit from a monogamous relationship, if their mate supports them or their offspring in some way, it is difficult to understand why males would tie themselves to one mate when mating is not costly for them. There are several theories regarding the adaptive significance of pairing in primates. First is the idea that the female needs help defending a territory in order to obtain enough resources for herself and her offspring. Couples may actively and passively defend their territories (hence the more appropriate term "territorial pair") via threats, fighting, and duetting, i.e. calling together to indicate that the territory is occupied by a bonded pair. In the majority of species, males help by carrying offspring. The second theory suggests that monogamy is a way for males to protect their offspring from infanticide. In those species that form one-male groups, when a new male takes over, he may kill nursing infants. Once nursing is interrupted, a female undergoes hormonal changes and may return to estrus (fertile period). It is in the new male's best interest to impregnate females as soon as possible, in the hope that some of his offspring will make it to the juvenile stage before the next male comes in and wipes out the infants. It is not in their best interest to wait to reproduce either. That is the way natural selection works. Those traits that maximize fitness, i.e. reproductive success, are favored. In addition, a male offspring that grows up to be infanticidal will be in a better position to reproduce, if he has what it takes to take over a group.

There are territorial paired species within the prosimians (indris and wooly lemurs of Madagascar and the tarsiers of the Southeast Asian archipelago), New World monkeys (night, titi, saki, and some marmoset monkeys of Central and South America), and the lesser apes (gibbons and siamangs of Southeast Asia). Females of some gibbon species tend toward polyandry and thus males are polygynous, making those species polygynandrous. We are the only great ape to have a tendency for monogamy, in that we tend to "fall in love" with one person at a time.

One-male Groups and Polygyny

In some species, one male with one or a few females is the grouping pattern. However in other species (Hamadryas baboons, geladas, mandrills, drills, and some odd-nosed monkeys, such as snub-nosed monkeys), one-male units (OMUs) congregate into larger and larger groupings, in a multi-tiered or nested fashion, depending on their current activity.. Except for the gorillas, all OMG species are Old World monkeys. The majority of the colobines form OMGs, e.g. African colobus monkeys of the genus: *Colobus* and Asian langurs and leaf monkeys. Cercopithecines of the genus: *Cercopithecus* (commonly known as guenons) and patas monkeys (*Erythrocebus*) are also OMG species.

De Brazza's monkey.

In the majority of OMG species, females are related but as groups get larger, they split along matrilines, meaning that a group of closely related females may splinter when competition increases. In addition, females may move between groups, especially in the colobines. Males fiercely compete for access to groups and infanticide occurs during takeovers. In those species that are seasonal breeders, it is difficult for the male to monitor and mate with all of the females and outsider males may sire some of the offspring. One guy can only do so much and females only have a small window of opportunity.

While the OMG makes sense for the colobines and their high leaf diet, it is not as clear why the more generalist guenons exhibit the same pattern. Like the colobine monkeys of Africa and Asia, it is possible that the ancestor of the extant arboreal guenons never left the trees and thus did not evolve the tendency for a larger grouping pattern in response to terrestrial predators. In addition, if they remained arboreal in relict forests, they may have enjoyed a more stable resource base. They are small- to medium-sized monkeys and thus can subsist on a variety of foods, primarily insects and fruit, both of which are indefensible food items, from a female perspective. Thus while a group is beneficial, it does not need to be large. It may be a bit of an oversimplification that female resources drive primate social organization, but it is a useful model with demonstrated heuristic value.

Hamadryas baboons.

For those species with a nested grouping pattern of OMUs, will describe the system in Hamadryas baboons and contrast it with geladas. Both species consist of OMUs that congregate into three larger group levels. For some strange reason (as if there are not enough terms in a primate course), some primatologists use different terms for the levels in each of the species. The basic unit is the OMU. The next level is termed the clan; it consists of several OMUs, along with bachelor males, and the members tend to forage together. The third level is the band, and that is the result of several clans congregating to forage over a large area. While Hamadryas bands are somewhat stable, gelada bands are not. Finally the troop (Hamadryas) or herd (gelada) is a combination of multiple bands that come together to sleep on cliffs in the mainly treeless regions where both species live, primarily in Ethiopia. Troops consist of hundreds of animals, over 700 in the Hamadryas and slightly fewer in the geladas. This odd grouping pattern is related to their harsh environment. Hamadryas live in subdesert conditions in Ethiopia and the Arabian Peninsula. They are generalists that eat whatever they can find. They fission and fuse (i.e. come together and separate again) into the various grouping levels as resources allow, but predators abound and shelter is scarce, so there is safety in numbers via vigilance. The geladas' situation is a bit different. They live in high-altitude conditions in Ethiopia and eat a lot of grass and grass products, such as seeds and corms. Again, there is safety in numbers but resources are ubiquitous so they spread out a bit and mooch and munch (new foraging category), i.e. sit-eat-move, along the ground. The strangest aspect of the two species is that Hamadryas are male philopatric and geladas are female philopatric. While the gelada pattern makes sense, considering their relatedness to female-philopatric baboon species, Hamadryas are even more closely related to those baboons yet appear to deviate from the pattern. However, the females do not go far; they transfer at the clan or band level and thus are not far from kin.

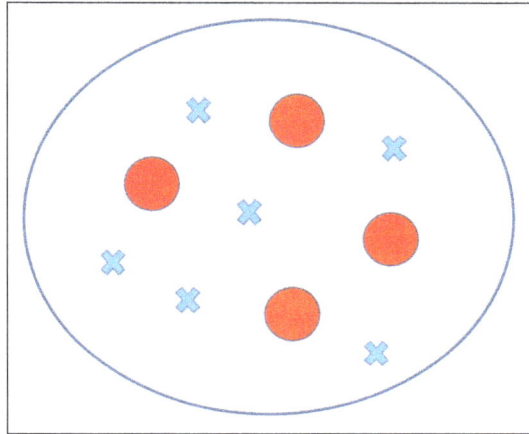

A depiction of a gelada or Hamadryas baboon clan.

Grazing geladas.

While there are regular takeovers in gelada OMUs, as would be expected, they are not as frequent in Hamadryas, primarily due to the facts that males have control over their small group of females and OMUs are surrounded by male relatives. Hamadryas females are usually coerced away from their mothers when they are young and then herded and punished by their new male leader until they learn to obey and not stray. Female geladas have a say in male takeovers; they either side with the resident male and help keep the new male out or they do not and the resident male is on his own. It is interesting that if a new male becomes established in the group, the former male may stay and help defend his offspring from becoming the victims of infanticide, but he can no longer breed.

One-female Groups and Polyandry

Emperor tamarin.

This type of social organization is seen only in the callitrichids, i.e. the tamarins and marmosets of Central and South America. Within those groups, there is usually only one breeding female and one or two breeding males. Females gestate as many as five fetuses but on average, only two survive. Hence we talk about twinning in the callitrichids. Those groups with an extra male have better offspring survival. At birth, the offspring average one-fourth of the female's weight and thus foraging to support them is a full-time job for the females. The females nurse the young and the males carry and nurture them.

Females pheromonally suppress cycling in their daughters and while sons become fertile, they have no mating options in the group. Mature daughters and sons also help with the care of their younger siblings. Helping behavior, while delaying their direct fitness (genes they pass on via reproduction), increases their inclusive fitness (genes they share with relatives). Full siblings share half of their genes (the same as between parents and offspring) and half siblings share one-fourth, on average. While the proximate causation (current stimulus or condition favoring the behavior, versus ultimate causation, i.e. the behavior was favored by natural selection due to its fitness benefits) for older siblings to stay is unclear, it is likely adaptive in some situations to delay reproduction. For example, it may be difficult for young animals to compete for territory or mates, and they are small and inexperienced and thus easy prey.

Multi-Male/Female Groups and Polygynandry

There are two types of multi-male/female groups (MMF). The first is the more common. They are medium to large groups of related females (female philopatric) with a sex ratio skewed in favor of females. Outsider males may congregate in all-male bands. Females and males are promiscuous, the mating pattern known as polygynandry. Many New World monkey species and most of the Old World cercopithecines (e.g. macaques) exhibit this type of social organization. Females cooperate in resource defense and males may have a more peripheral position within the group, except during the mating season in seasonal breeders. Many semi-terrestrial species exhibit this type of social organization, e.g. baboons (Hamadryas are the exception) and macaques. Terrestriality is associated with larger body and group size, likely for predation avoidance. With more females, come more males and with more males, females can benefit from seasonal breeding. There are enough males to go around and the glut of offspring that are then born reduces the probability that any one of them is eaten; i.e. the dilution effect. In addition, related females help keep watch over the young that then have playmates. Seasonal breeding is tied to environmental conditions, so that females benefit by timing events to coordinate with resource availability.

Long-tailed (also known as crab-eating) macaque. "Ngarai Sianok Sumatran monkey".

The second type of MMF is commonly called a community social organization. Species that exhibit this type of social organization are male philopatric ripe fruit specialists. females cannot defend fruit, so they do not band together into matrilines. Related males defend a territory that contains enough resources to attract females. Females and their offspring forage independently but group members come together periodically into larger aggregations, hence the other term for this type of social organization, fission-fusion. New World spider and muriqui monkeys and the chimps and bonobos of Africa are all categorized as community species.

Bonobo group hug.

Dominance

Primates, as (mostly) group living animals tend to form what are known as "dominance hierarchies". Animals higher in the hierarchy tend to displace lower ranked individuals from resources (mates, space, food). They tend to have higher reproductive success (either by mating more often, or by having more resources to invest in their offspring). The hierarchy is not fixed and depends on a number of changing factors (age, sex, aggression, intelligence perhaps), and may also depend on the support of others.

The rank is learned through play, agonistic interactions and affiliative interactions (and rather tautological, that's exactly how it's measured too). This maintenance of social position, and social knowledge of ones rank is one of the postulated theories for why humans have been forced to evolve large brains.

Grooming

This is a common primate activity. Allogrooming (others) is an important affiliative mechanism. It can be used to strengthen links, subordinate animals tend to groom more dominate ones; males groom females for sexual access. Or for more practical purposes: mothers grooming infants to keep their fur clean. It is certainly the cement that keeps the primate social structure together.

Communication

This includes scents; body postures; gestures; vocalisations. Some of these appear to be autonomic responses indicating emotional states: fear, excitement, confidence, anger. Others seem to have a more specific purpose: loud ranging calls in Indri, howler monkeys and gibbons; quiet contact calls in lemurs to keep the group together; fear calls in lost infants, or on spotting predators. From our human perspective, we often find it easier to associate sounds with specific meaning, but among NHPs, gestures and actions are often used. Presentation and mounting behaviour is often used to diffuse potentially aggressive situations. Yawns exposing teeth are often threats, as is direct eye contact. Apparently this can cause problems when looking at baboons with binoculars: the front lenses look like bigger than normal eyes and this is seen as the observer being very aggressive.

Facial expression is important too. It's very obvious in chimps: their expression often appear all to human-like; but other primates also use stereotyped eyelid flashes or lip slaps.

In addition, there has recently been a great deal of success teaching chimps human language. This was initially American sign language, but has now been extended through the use of modified computer keyboards to really very high levels of sophistication (especially Kanzi, a pygmy chimp).

Mothers and Infants

This is the basic social group for many primates. It has been observed that this mother infant bonding is needed to allow the infant to be able to interact properly as an adult, and, if female, to be able to cope with offspring. This is one of the big problems with zoo animals where an individual has been hand reared by keepers. In some primates, this mother infant closeness continues after infancy. The females remaining in the group as a matriline and the males dispersing to other groups. The combined power of one of these female bonded matrilines is enough for the group of females to be more dominant than the alpha male, even though he is much bigger than an individual member.

Aggressive and Affiliative Behavior

Many behaviours exist to keep the group structure running smoothly for the members of the group. There are occasions though when these behaviours (especially aggression) are directed outside the group. Baboons gang up to repel attacks by hyenas, and chimps have been known to systematically gang up on and destroy neighbouring groups of chimps.

One interesting argument here, is that the development of bipedalism has been seen by some to be driven by a root as an aggressive, dominance display behaviour. This is the gorilla standing bipedally and banging his chest, or a male chimp bipedally charging a subordinate. Most people would probably consider this to be an effect of a bipedal ability, rather than the cause.

Cultural Behavior

This is learned behaviour that is passed from generation to generation. You will hear a lot about this in humans, but it has been observed in primates too. One prime example is a group of Japanese macaques, where one individual accidentally learned that the sweet potatoes that they were being fed tasted better if the sand was washed off (this is not a normal food for these animals). This behaviour has now spread through the whole group, and is being passed on to infants. It is now part of their culture. Tool use abilities are often thought to be acquired and passed on in this way too - for example, termite fishing in chimps.

Cetacea Behavior

Sleep

Conscious breathing cetaceans sleep but cannot afford to be unconscious for long, because they may drown. While knowledge of sleep in wild cetaceans is limited, toothed cetaceans in captivity have been recorded to exhibit unihemispheric slow-wave sleep (USWS), which means they sleep with one side of their brain at a time, so that they may swim, breathe consciously and avoid both predators and social contact during their period of rest.

A 2008 study found that sperm whales sleep in vertical postures just under the surface in passive shallow 'drift-dives', generally during the day, during which whales do not respond to passing vessels unless they are in contact, leading to the suggestion that whales possibly sleep during such dives.

Diving

While diving, the animals reduce their oxygen consumption by lowering the heart activity and blood circulation; individual organs receive no oxygen during this time. Some rorquals can dive for up to 40 minutes, sperm whales between 60 and 90 minutes and bottlenose whales for two hours. Diving depths average about 100 m (330 ft). Species such as sperm whales can dive to 3,000 m (9,800 ft), although more commonly 1,200 metres (3,900 ft).

Social Relations

Most whales are social animals, although a few species live in pairs or are solitary. A group, known as a pod, usually consists of ten to fifty animals, but on occasion, such as mass availability of food or during mating season, groups may encompass more than one thousand individuals. Inter-species socialization can occur.

Pods have a fixed hierarchy, with the priority positions determined by biting, pushing or ramming. The behavior in the group is aggressive only in situations of stress such as lack of food, but usually it is peaceful. Contact swimming, mutual fondling and nudging are common. The playful behavior of the animals, which is manifested in air jumps, somersaults, surfing, or fin hitting, occurs more often than not in smaller cetaceans, such as dolphins and porpoises.

Whale Song

Males in some baleen species communicate via whale song, sequences of high pitched sounds. These songs can be heard for hundreds of kilometers. Each population generally shares a distinct song, which evolves over time. Sometimes, an individual can be identified by its distinctive vocals, such as the 52-hertz whale that sings at a higher frequency than other whales. Some individuals are capable of generating over 600 distinct sounds. In baleen species such as humpbacks, blues and fins, male-specific song is believed to be used to attract and display fitness to females.

Hunting

Pod groups also hunt, often with other species. Many species of dolphins accompany large tunas on hunting expeditions, following large schools of fish. The killer whale hunts in pods and targets belugas and even larger whales. Humpback whales, among others, form in collaboration bubble carpets to herd krill or plankton into bait balls before lunging at them.

Intelligence

Bubble net feeding.

Cetacea are known to teach, learn, cooperate, scheme and grieve.

Smaller cetaceans, such as dolphins and porpoises, engage in complex play behavior, including such things as producing stable underwater toroidal air-core vortex rings or "bubble rings". The two main methods of bubble ring production are rapid puffing of air into the water and allowing it to rise to the surface, forming a ring, or swimming repeatedly in a circle and then stopping to inject air into the helical vortex currents thus formed. They also appear to enjoy biting the vortex rings, so that they burst into many separate bubbles and then rise quickly to the surface. Whales produce bubble nets to aid in herding prey.

Killer whale porpoising.

Larger whales are also thought to engage in play. The southern right whale elevates its tail fluke above the water, remaining in the same position for a considerable time. This is known as "sailing". It appears to be a form of play and is most commonly seen off the coast of Argentina and South Africa. Humpback whales also display this behaviour.

Self-awareness appears to be a sign of abstract thinking. Self-awareness, although not well-defined, is believed to be a precursor to more advanced processes such as metacognitive reasoning (thinking about thinking) that humans exploit. Cetaceans appear to possess self-awareness. The most widely used test for self-awareness in animals is the mirror test, in which a temporary dye is placed on an animal's body and the animal is then presented with a mirror. Researchers then explore whether the animal shows signs of self-recognition.

Critics claim that the results of these tests are susceptible to the Clever Hans effect. This test is much less definitive than when used for primates. Primates can touch the mark or the mirror, while cetaceans cannot, making their alleged self-recognition behavior less certain. Skeptics argue that behaviors said to identify self-awareness resemble existing social behaviors, so researchers could be misinterpreting self-awareness for social responses. Advocates counter that the behaviors are different from normal responses to another individual. Cetaceans show less definitive behavior of self-awareness, because they have no pointing ability.

In 1995, Marten and Psarakos used video to test dolphin self-awareness. They showed dolphins real-time footage of themselves, recorded footage and another dolphin. They concluded that their evidence suggested self-awareness rather than social behavior. While this particular study has not been replicated, dolphins later passed the mirror test.

Horse Behavior

Horses have unique and fascinating behavioral characteristics which have contributed to their development, survival and present-day value as a companion to people. The successful 4-H Horse Project member should learn to understand horse behavior, and apply this knowledge to all facets of interaction with horses.

Ethology is the scientific study of animal behavior. Technically, ethology is the study of animals in their natural habitat, but most behaviorists agree that the behavioral traits of domestic horses are relatively similar to horses before domestication.

Behavior can be defined as the animal's response to its environment. Because domestic horses exist in a relatively controlled environment, their response is fairly predictable.

Several things make a horse unique in the animal kingdom.

- Horses are strongly social. They are herd animals, which are at a higher comfort level when they maintain a visual contact with other horses.

- Horses are herbivores. They eat plants. They rely on grazing grasses and leaves for feed.

- Horses typically show a tendency for imitation between young and old.

- Horses are seasonal breeders and, as a consequence, foaling patterns occur.

- Males tend to form a separate male sub-group structure at certain times of the year.

- Horses are considered prey species within the animal kingdom.

- Horses are capable of strong pairbond relationships.

Much of what makes a horse behaviorally unique is related to being a herbivore (plant-eating animal) and a prey species. To understand this, consider a how horse's feeding behavior differs from a carnivore's (meat-eating animal) feeding behavior. Carnivores spend a greater proportion of their time stalking food and less time consuming food. Horses spend a greater proportion of their time consuming food and less time searching for food. Carnivores typically must attack and subdue their food before consuming it; horses merely graze and browse. These and many other feeding behavioral traits can be logically assumed to provide a basis for much of what makes horses different from carnivores. Carnivores are aggressive, horses passive. Carnivores are anatomically suited for killing other animals; horses are anatomically ideal for grazing and browsing. Carnivores can kill horses. Horses are less likely to kill carnivores and are more likely to try to escape from a carnivorous threat.

Types of Behavior

Animal behaviorists have classified the social behavior of horses (and other animals) into the following categories:

- Contactual Behavior - Behavior related to seeking affection, protection or other benefits by contact with other animals. Communication behavior is sometimes considered as a separate category.

- Ingestive Behavior - Behavioral activities associated with eating and drinking.

- Eliminative Behavior - Behavioral activities associated with defecation and urination.

- Sexual Behavior - Behavior related to mating between males and females.

- Epimeletic Behavior - Behavior related to giving care and attention, most common between a mare and foal, but also between other horses, such as horses standing together under shade and "switching" flies from one another.

- Allelomimetic Behavior - Behavior related to mimicry; contagious or infectious behavior such as when one horse copies the behavior of another. If one horse starts running, for example, others are likely to join in. This may be a defense maneuver that is typical of wild horses.

- Investigative Behavior - Behavioral activities associated with curiosity; the exploration of the surroundings or objects. Horses are noted for using all their senses to thoroughly "check out" any new item, horse or place with which they are presented.

- Agonistic Behavior - Behavior associated with conflict or fighting, including anger, aggression, submission and flight from conflict. Sometimes behaviorists separate this into two categories (aggression and fearfulness).

- Dominance\Submission - Behavioral activities often referred to as "pecking order," because the early behavioral work in this area was done with poultry. Dominance hierarchies are extremely prevalent in the social order of horses.

Dominance is generally established through agonistic behavior, and may be extremely violent (such as fighting between stallions) or as simple as threatening looks (ear pinned back, squeals, sudden moves in the direction of the submissive animal). If the lower-ranked (submissive) animal has room to escape, there will often be no contact, and the hierarchy is there- fore established or maintained with little or no fighting.

Horse Senses

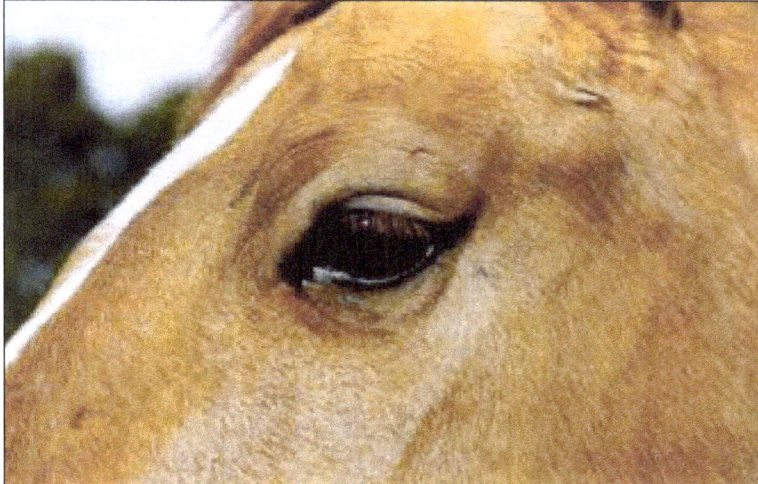

Horses have a very large eye and a very large pupil.

The senses are an important part of what makes horses behaviorally distinct. Animals share the five basic senses: vision, audition (hearing), olfaction (smell), gusta- tion (taste) and touch. The senses are the tools that an animal uses to interact with its environment. As such, the senses can be considered starters of behavior.

There is a temptation to relate human senses to horses, but horses and people have basic differences in how they see, feel, taste, smell and hear their environment. We do not completely understand horse senses, but the things we have learned have greatly added to our horse knowledge. A review of this information can be helpful in understanding horses.

Vision

Ever look at a horse eye to eye, you probably noticed a few things. First, they have a very large eye and a very large pupil. Second, the eyeball is placed more to the side of the head, which gives horses a wider field of vision.

Predator species, such as dogs and coyotes, have eyes placed toward the front of their head. This narrows their total field of vision but it increases their binocular (using two-eyes) visual field. Binocular vision gives the predators better depth perception and a more concentrated field of vision. Prey species, such as horses, sheep and cattle, have a much wider visual field. With only slight head movement, horses can scan their entire surroundings. If there is a threat, the behavioral response is generally to flee.

Much of the width of the visual field that horses see is observed with only one eye.

This is called monocular vision. When a horse sees an object with its monocular vision, it will tend to turn toward it so that both eyes can see it (with binocular vision), and the ears can better hear it. There is sometimes a brief visual shift as the horse switches from monocular to binocular vision, which sometimes causes an unexplained spooking of the horse.

The size of the pupil improves the ability of a horse to pick up movement. The large size provides a built-in wide angle lens effect which is further enhanced by the placement of the visual receptors in the retina. The total effect is better side (peripheral) vision. The horse can see movement very well.

Does the horse sacrifice visual accuracy to get a wider field of vision? In general, yes, but the answer to the question is not clear. Current thought is that, while the horse sees practically all the way around its body, the image is not as clearly defined as what humans see, especially within four feet. This, plus the fact that a horse cannot see directly below its head, may explain why horses often raise their head to observe close objects. Conversely, a horse tends to lower its head to observe faraway objects.

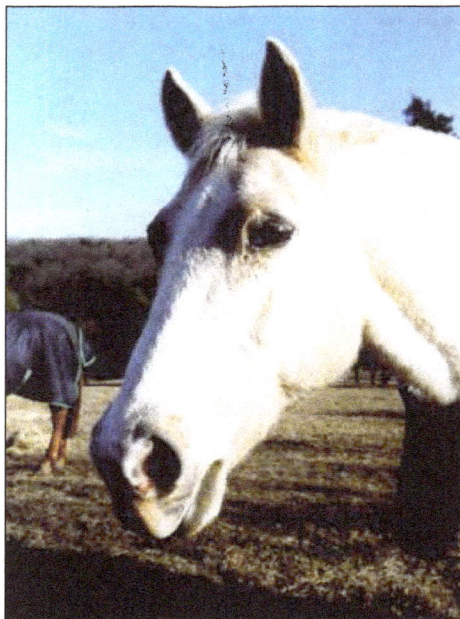

A horse will turn toward an object so that both eyes can see it (binocular vision).

In spite of the wide field of vision, there is a "blind spot" directly behind the horse. People should avoid approaching a horse from behind, because their presence may not be detected until they are close, and this could startle the horse. Some horses may instinctively kick in this situation. If approaching a horse from the rear cannot be avoided, make a soothing noise to announce your presence.

Do not "sneak up" on a horse from behind.

Another question often asked is do horses have color vision? For many years it was believed that both horses and cattle were color blind. If horses can distinguish colors, it is unlikely that horses' ability to see color is equal to other species, such as humans.

Hearing

In spite of its importance, there is limited information about the auditory (hearing) sense of horses. We know horses are sensitive to high-pitched noises and the release of stress-related hormones in response to sudden loud noises such as firecrackers or barking dogs.

Horses become nervous and difficult to handle when stress hormones are elevated, so it may be useful to avoid loud or shrill noises when handling or moving horses.

The horse can amplify and pinpoint sound with its ears. Sound arrives at each ear at slightly different times, which allows the horse to use sound as a means to tell where the sound came from.

The horse can then move its ears, head or its entire body to tell more about the source of the sound. This skill is probably as important as sight and smell for keeping the horse, as a prey species, alive.

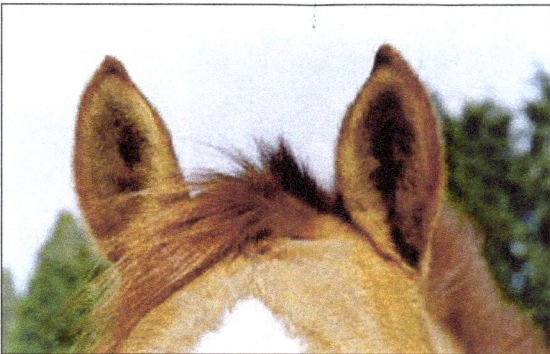

A horse can amplify and pinpoint sound with its ears.

A horse can rotate its ears independently from front to side to pinpoint a sound.

Smell

The horse's sense of smell (olfactory) may be the most difficult for humans to understand.

Horses have a more highly developed sense of smell than humans, and they use their ability to distinguish different odors more in their everyday lives.

Horses use their sense of smell in a number of ways. Horses use smell to identify other horses, particularly when a mare uses smell to pick out her foal from a group.

Another common use of smell is during mating. The stallion constantly checks mares to detect the ones in heat (estrus). The classic head-raised, lip-curling behavior of the stallion (bulls and rams, also) as he smells females is called the Flehmen response. This trait, which may be occasionally observed in females, is due to a special organ (vomeronasal organ) above the roof of the mouth, which humans do not have.

Horses probably use their olfactory sense to locate water and identify subtle or major differences between pastures and feeds. Smell also triggers behavioral responses. There are, for example, horses that do not like the smell of tobacco smoke or may react negatively to the odor of certain medications.

Some people believe that horses can sense when a person is afraid which is probably true and this is often referred to as horses' ability to smell fear. It is possible that the horse can smell some small change in the fearful human, but it is equally likely that the horse can sense the human nervousness via other senses.

The raised-head, lip-curling behavior usually displayed by stallions during mating is called the Flehmen response.

Horses will use their sense of smell to select fresh feed in preference to spoiled feed. The next time you are tempted to dispose of moldy feed or hay by feeding it to a horse, try smelling it yourself. Then remember, if it smells bad to you, it may smell worse to the horse. (This may not always work, however, because some molds, such as highly poisonous aflatoxins, cannot be detected by humans).

Taste

The sense of taste in horses is probably not as important as the sense of smell, and it is difficult to separate behavioral responses that are due primarily to taste from responses caused by the olfactory sense. Using their sense of taste, however, is part of why horses can tell one feed from another. When presented with a variety of feeds, horses will select certain feeds over others. In practical situations, such as under grazing conditions with multiple forage species present, the horse will select different types and species than either sheep, goats or cattle.

There have been experiments to determine if animals have nutritional wisdom. This is based on the premise that horses will attempt to eat feeds that provide them with the nutrients needed. In most cases, however, horses are unlikely to balance their own ration when provided a variety of feeds. If possible, they will consume feeds at a level far higher than necessary to provide essential nutrients. For example, salt is often provided to meet horses' requirement for sodium; however, horses will often consume many times the amount of salt needed to meet the requirement. Fortunately, there is no evidence that over consumption of salt will cause health problems if adequate water is available.

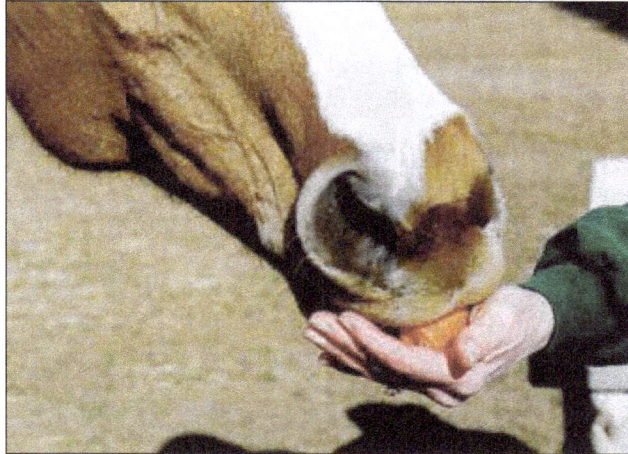

Horses enjoy special treats, such as carrots or alfalfa cubes.

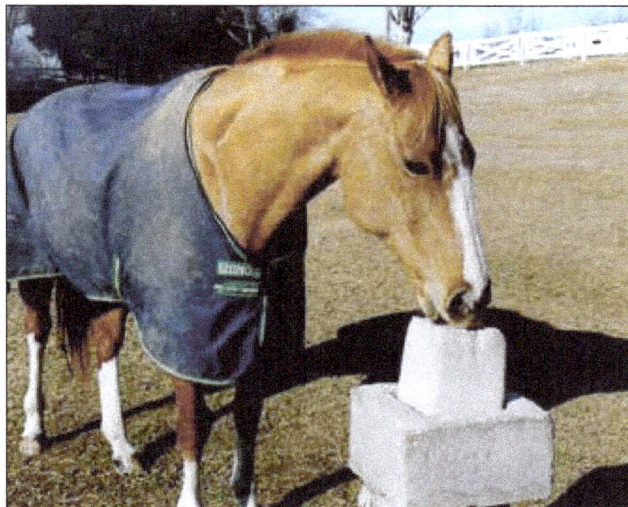

Placing a salt-lick in a paddock will help a horse satisfy its sodium requirements.

Touch

The sense of touch is certainly well developed in horses, and is one of the most important senses in terms of human interaction with horses. The nose, lips, mouth and possibly the ears are the most sensitive areas to touch and, consequently, most readily lend them selves to feeling behavior. Although hooves do not respond to touching, they should not be regarded as without feeling. In fact, various parts of the hoof are able to feel touch, as anyone who has shod horses or trimmed hooves can relate.

Other areas of the body are also sensitive to touch. The flanks for example, are particularly sensitive, and can pick up a light signal from the rider. The ribs are also sensitive, as are the withers and back.

Understanding the degree to which horses are sensitive to touch can be valuable to the trainer. For example, knowing that horses can feel the slightest touch with their lips underscores the importance of developing a light touch on the reins, and making certain that bridles be correctly fit to the horses head and mouth. Knowing that the horse can feel the slightest shift of weight in the saddle illustrates why the rider's position is important as the mount is guided toward a jump or other maneuver. Poor position, exaggerated movement or excessive force are confusing to horses and result in poor performance.

The sense of touch is undoubtedly important in interaction between animals. Foals seek bodily contact with their dams (mothers), and mares respond to the touching behavior of their foals in various ways, including milk let-down in response to the nuzzling/suckling stimulus of foals.

Another example of horses' sensitivity to touch is related to electric fences. Anyone who has used electric fences with a variety of grazing animal species knows that horses are very sensitive to electricity. To use electric fences with horses, the wire should be placed approximately at nose height. High-quality, well-grounded chargers should be used, and horses should be trained to the fence by introducing them to a well-constructed permanent electric fence for their first experience.

A rider's position is very important, because horses can feel the slightest shift of weight in the saddle or pressure from the rider's leg.

The Role of the Senses in Training

The horse must rely on its senses in order to perceive the signals (often called cues) that the rider is giving. Touch and sound are the primary senses which are used.

This is not a horse training manual. There are a number of training publications, often developed by breed organizations or successful trainers, which can provide more detailed information about how to train your horse.

However, understanding the behavioral basis that the horse has for recognizing cues through its senses can be helpful in training.

The basic steps for using senses in training are:

- Stimuli - The trainer/rider initiates a cue, thus providing a stimulus to the horse.

- Sense - The horse senses the stimulus.

- Response - The horse responds to the stimulus with an action.

- Reinforcement - The trainer reinforces in a positive way by rewarding the correct response and in a negative way by discouraging an incorrect response.

Good trainers recognize that each horse has its own combination and will develop at its own pace. Intelligence, individual energy level, previous experience and many other factors may affect response. Patience, repetition and building in small increments of success will give the best results. Over-use of negative reinforcement may yield a horse that is prone to nervousness.

Use positive reinforcement more than negative if long-term development is desired. Do not expect reasoning powers that are beyond the powers of the horse to give.

An extremely well-trained horse that was trained by a professional is likely to "come untrained" when ridden by a novice if reinforcement schedules are not maintained.

Horse Communication

Horses communicate in many ways, including visual displays, sounds and even through smell. Understanding how a horse communicates is important to the 4-H Horse Project member for three important reasons:

- Diagnosis of medical problems - Learning how a horse behaves when it is sick is important not only so you will know it needs attention, but also because certain behaviors are linked with specific problems. Naturally, a veterinarian or other experienced person may need to be called upon for more detailed information, but it is never too soon for a young person to start learning to tell the difference between normal, healthy behavior and the behavior of sick animals.

- Assessment of Emotional State or Temperament - As your experience and skill in assessing horse behavior increases, you will find that you can read a horse's emotional state. This knowledge can be applied when assessing how friendly a new or unknown horse is, or it can be used to tell if a horse you ride every day is in an unusual mood or is experiencing emotional difficulty. Sometimes a horse is reacting to the presence of another animal, or it can be related to something as simple as the weather.

- Safety - Horses usually communicate a warning before they cause harm. Learning the communication signals ears back, head lowered, teeth bared, turning into a kicking position, tail swishing, etc. that warn of danger are important in avoiding harm. Nervous or "jumpy" behavior can possibly be as dangerous as aggressiveness. Learn to recognize the signals that could result in harm, whether it comes from the horse you are riding or one that is being ridden by someone else in your vicinity.

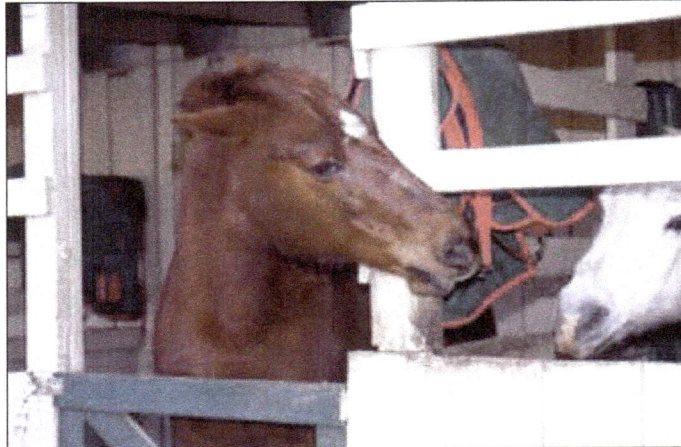

Horses can often display aggressive behavior when being fed.

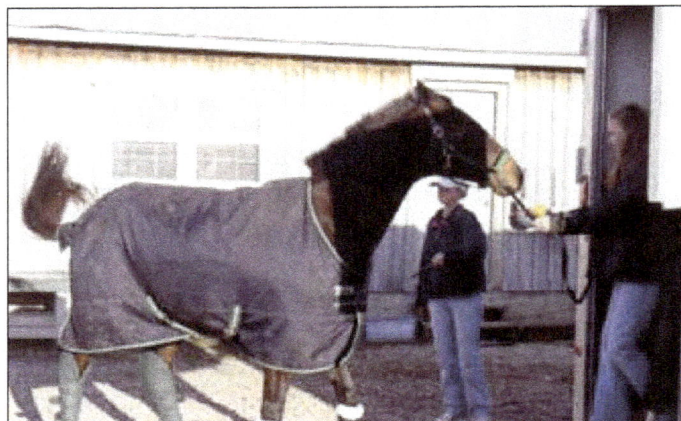

Unfamiliar situations such as boarding a trailer can make horses jumpy and nervous.

Mating Behavior

Behavior is an important aspect of reproduction, and an understanding of the basics of reproductive behavior can lead to management applications that can improve reproductive success.

Puberty

Puberty is the attainment of sexual maturity. In fillies, this can be as early as nine or 10 months, but is usually 12 to 15 months.

Stallions are 15 months or older before they can successfully breed. Behaviorists have noted that both stallions and fillies (less frequently) may exhibit sexual display before the reproductive tract is physiologically mature. Therefore, pregnancy cannot occur. Conversely, some fillies may cycle, but not exhibit estrus.

Estrus (Heat)

Estrus, or heat, is the period of the reproductive cycle when the mare ovulates and, if bred, is likely to conceive. A behaviorist would define estrus as the behavioral state when the female seeks and accepts the male.

The average length of the estrus cycle, or the period from heat to heat, is 21 days, but can vary from 19 to 26 days. The duration of the estrus period is typically a week (actually about 6 days), but varies from two to 10 days. The foal heat, or postpartum estrus, typically occurs six to nine days after foaling, but may be as early as five days or as late as 15.

It is important for the mare owner to recognize the behavioral signs of estrus. Some signs are general, including restlessness, hyperactivity, less time devoted to eating and resting and more time "running the fences." Other signs, such as frequent urination, straddling (squatting) posture and clitoral "winking" are more specific and are often not as obvious in early stages of estrus.

Some mares are more likely than others to exhibit overt signs of estrus. Older and more experienced mares are more likely to exhibit clear signs of estrus. Maiden mares are considerably more likely to cycle without visible signs of heat.

The presence of stallions increases the behavioral display of estrus in mares. People who keep only one or a few mares often have difficulty in identifying the onset of estrus. Sometimes such small operations may benefit from having a single stallion (sometimes a pony is preferred) around as a "teaser," to stimulate estrus display. The use of a teaser stallion in larger breeding operations is routine to stimulate mares into a receptive state before the introduction of the breeding stallion.

Seasonal Breeding Behavior

Horses exhibit seasonal breeding patterns. In general, they are referred to as "long-day breeders," because as the days increase in length in the spring, they come into heat. Mares are also called "season ally polyestrous" because they have multiple cycling periods. The most likely breeding season for horses is the spring or summer. Since light is a factor in controlling the seasonal breeding pattern, horses are some times called "increasing-light" breeders. Most studies have indicated a tendency toward anestrus (not cycling) in the winter months; however, some mares may cycle during this time as well.

Courtship and Mating

Mares will cycle several times during the breeding season if they are not bred or if they fail to become pregnant. The heat period (mentioned previously) is about a week, but the most intense estrus behavior, when the mare is most sexually receptive to the stallion, is about three days.

Mares in heat may actively seek out and attempt to stay in the vicinity of the stallions. There may be few other signs of estrus early in the heat period. As the heat period progresses, the mare may become more active in her courtship behavior. During the peak of estrus, the mare may sniff, lick or nuzzle the stallion. A mare in heat is likely to urinate frequently, particularly if a stallion is investigating. She is also likely to raise her tail and assume a breeding stance. A mare may exhibit the urination response to the stallion whether she is in heat or not; however, if she is not in heat, she will usually leave the vicinity of the stallion or turn on him with threat behavior. If she is in heat, she may passively accept the attention of the stallion, occasionally turning her head to observe the stallion's activities. The classic behavioral display of the stallion when it checks a mare is to lift its nose into air and curl its upper lip. This is called the "Flehmen" response.

Stallions exhibit certain additional stereotypical display patterns. They will often be impatient, alert, hyperactive and restless. Vocalization is common. The stallion will frequently nudge the mare, apparently to signal readiness and to assess her "firm stance" response. In addition to nudges, some stallions may smell and bite over the mare's body. Most behaviorists consider the display as being more important in the courtship process than odor recognition.

Dominance Effects

Dominance patterns are very much a part of horse breeding behavior, particularly in natural environments. One stallion will typically dominate the breeding of a band of mares and competing stallions will be banished to form their own separate band until one of them become old enough, brave enough or tough enough to defeat the dominant stallion.

In modern breeding establishments with numerous, separately stalled breeding stallions, all the stallions are used for breeding.

Dominance, nevertheless is in evidence. Most breeding barn managers can tell you which stallion is dominant or "the boss."

Libido

Libido is the term which is used to denote sexual drive or the degree of sexual urge of animals. A stallion with a high libido will exhibit an eagerness to mount and attempt to breed a mare. In natural situations, stallions exhibit a wide range of libido levels, from zero activity to the extremely aggressive stallion who sacrifices all other pursuits in favor of searching for and breeding estrus females. Either extreme may cause problems, and young stallions are more likely to exhibit extremely low or high libido.

Feeding Behavior

The behavioral definition of a horse includes reference to the fact that the equine is a grazing animal. In other words, the way the horse eats is an integral part of what makes a horse a horse. Since the horse is an ungulant, it is predictable that much of the behavior that is demonstrated is related to the consumption of forages.

Behavior has direct effects on consumption patterns, feed availability and the selection of feeds. Horses devote more time to eating than to any other behavioral activity. Feed and forage are generally expensive items in the budget.

There is probably no other single factor as important to the well-being and productivity of the horse than the feed and forage it consumes.

Basic Feeding Behavior

Horses have to consume feed and water in order to survive.

Beyond survival, consumption of the correct amounts and proportions of feedstuffs allow horses to thrive and be productive.

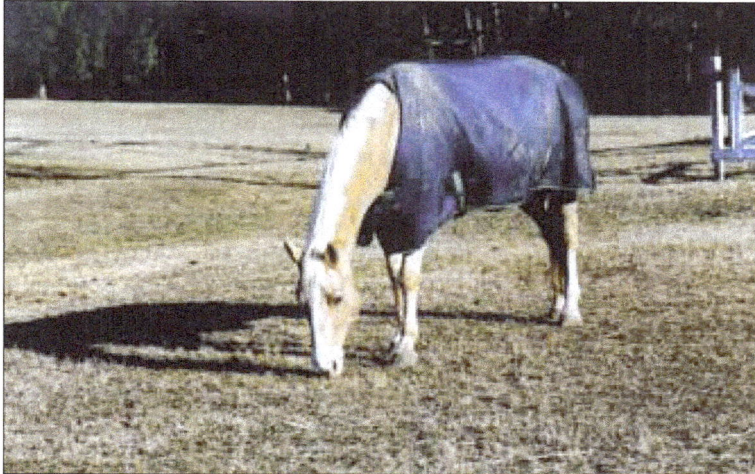

Horses spend approximately five to 10 hours per day grazing.

Relationship between Nutrition and Behavior

Feed consumption is motivated by hunger, but the methods and patterns of feeding are governed by behavior. Mare nutrient requirements, for example, increase during late pregnancy and lactation; therefore the demand for consumption of feed is higher (assuming there is no change on nutrient density of the ration). Horses also demonstrate increased appetite when work load increases. The horse compensates for this increase in demand by increasing the rate of eating.

If the quality or quantity of available feed is low or horses are being worked hard, the horse often cannot increase the rate of consumption enough to meet demand. This is where human management plays an important role. The owner should compensate for the imbalance and increase the feed and improve the feed quality.

It would be a mistake to oversimplify the relationship between nutrition and behavior. For example, there have been numerous attempts over the years to attribute "nutritional wisdom" to horses and other livestock. This is based on the idea that a horse will select a diet that is more nutritionally correct.

Unfortunately, this is often not the case. The horse selects a diet based upon a variety of factors, with nutritional value being of little importance in the food selection process.

Time Allotment for Ingestive Behavior

The time that horses spend consuming feed is governed by a number of factors. Grazing time depends primarily on: (1) type and availability of forage, (2) consumption behavior, and (3) the level of nutrient demand.

If feed is limited during periods of drought or when horses are fed a restricted feed allowance the horse will eat when feed is present or can be found. When abundant feed is available, horses will develop patterns of consumption behavior. These patterns of eating are developed in response to daylight/darkness cycles and other environmental cycles, and are apparently influenced by learned behavior as the horse grows and develops.

Most studies indicate that heaviest grazing will occur in the hours around dawn and in the late afternoon, near sunset. Night grazing also has been observed, and is likely to increase in the summer. Temperature can alter grazing times. When daylight temperatures become extremely warm, horses start and stop their grazing earlier in the morning on hot days. Cold weather alone apparently has little effect on daily grazing patterns; however, heavy rain, strong wind and snow cover may significantly alter grazing patterns.

Most estimates of time spent grazing fall between five and 10 hours per day. In general, horses spend less time grazing good quality pasture, but this is not always true. For example, although horses may graze poor quality pasture longer to meet nutritional requirements, horses on high quality pastures may consume forage for much longer than is necessary to meet needs. Over- grazing can lead the problem of horses becoming over-conditioned (fat) on pasture, because they are consuming more than they need to meet their nutrient requirements.

The problem of over-weight horses is most likely to occur when nutritional demands are lowest, such as in early pregnancy or when a working or show horse is turned out to pasture. It may be necessary to restrict access to pasture if horses are becoming over-conditioned.

Other Factors Influence Grazing Time

Feeding concentrate supplements, for example, may reduce forage consumption. Thin horses may consume more than fat horses, and this is at least partially explained by increased grazing time.

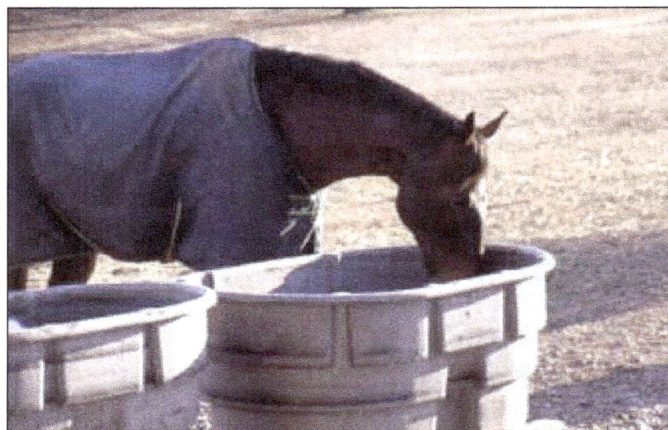

Horses need clean water available to them at all times.

Typically, horses do not drink water frequently hence the old saying "you can lead a horse to water but you can't make it drink." In natural environments, grazing patterns are often set to allow access to water once or twice per day. When horses do drink, they tend to take in a considerable amount, often in some 15 to 20 large swallows.

Selectivity

Horses have a relatively large mouth and remarkably flexible lips. They typically harvest the portion of the pasture plant they are interested in consuming by biting it off between their upper and lower incisors. They are able to graze close to the ground and are also able to comfortably adapt to browsing picking the leafy material from bushes or other plants. These anatomical/behavioral combinations result in the ability of horses to be selective about what they consume. The horse will often select the most tasty part of the hay and leave the stems and undesirable portions.

In the winter it is important to provide supplemental forages, such as alfalfa for good nutrition.

If an abundance of pasture is available, horses will be very selective. As the amount of avail- able forage decreases, the degree of selectivity will decrease. If a number of different varieties of forages are available, horses are more likely to demonstrate marked selectivity. However, if only one forage is available or if there are only a few species available and these are of similar acceptability to the horses, there will be little selectivity.

It is interesting that some excellent forages are not preferred by horses, and may be the last to be selected when others are available. One such example is alfalfa. Many reports indicate that when horses are provided a choice between grass and alfalfa, they will often select the grass first, even though alfalfa is nutritionally superior. However, when other forages are exhausted, the horse will quickly adapt to the alfalfa. Some researchers speculate that boredom or a desire for change is the reason why horses are occasionally observed selecting clearly inferior forage in the presence of abundant superior forage.

Sight, touch, taste and smell are used by the horse in selecting the forage species it will consume.

Taste is the sense that is most likely to influence selection. Indications are that odor plays a relatively minor role. Sight is probably used primarily to recognize conspicuous forage species and to orient the approach to those species, but sight apparently is not important in influencing selectivity. Horses eat leaves in preference to stems and green, succulent material in preference to dry, coarse material. Hunger tends to decrease selectivity.

Grazing Patterns

A number of factors can affect the grazing pattern. The location of water, for example, can have an important effect on grazing pat- terns. In arid zones, the water source is the center of grazing activity and the primary determinant of grazing the grazing area.

The area near the water may become overgrazed, even damaged and eroded, because of the influence of the water source on grazing pattern. Social factors, such as the development of a home or territorial area can inhibit movement of horses on large ranges. The social rank of the horses can determine which horses obtain the choicest grazing sites or best access to supplemental feed or water.

Cribbing (Horse)

Stereotypies are repetitive, unwavering behaviours that cease to obtain a goal and lack function. One of the most common stereotypies in horses is equine oral stereotypic behaviour, otherwise known as cribbing, wind sucking or crib-biting. Cribbing or crib biting involves a horse grasping a solid object such as the stall door or fence rail with its incisor teeth, then arching its neck, and contracting the lower neck muscles to retract the larynx. This coincides with an in-rush of air into the oesophagus producing the characteristic cribbing grunt. Usually, air is not swallowed but returns to the pharynx. Wind-sucking is a related behavior whereby the horse arches its neck and sucks air into the windpipe but does so without grasping an object. Wind-sucking is thought to form part of the mechanism of cribbing, rather than being defined as an entirely separate behavior.

Cribbing is considered to be an abnormal, compulsive behavior or stereotypy seen in some horses, and is often labelled a stable vice. The major factors that cause cribbing include stress, stable management, genetic and gastrointestinal irritability.

A similar but unrelated behavior, wood-chewing or lignophagia, is another undesirable habit observed in horses, but it does not involve sucking in air; the horse simply gnaws on wood rails or boards as if they were food.

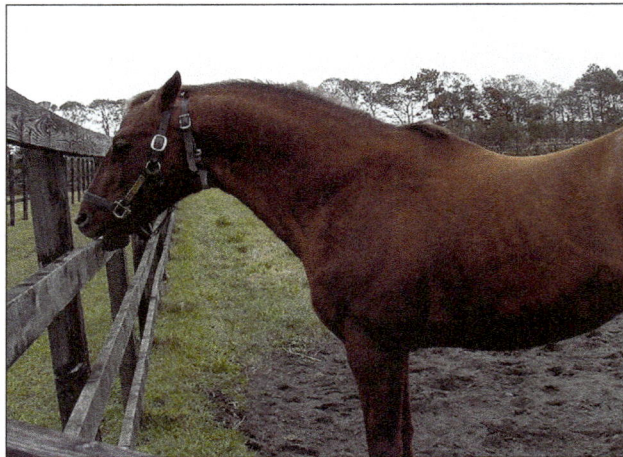

A horse cribbing on a wooden fence, note anti-cribbing collar intended to reduce this behavior and tension in neck muscles.

Cribbing, or crib biting, involves a horse grasping a solid object such as the stall door or fence rail with its incisor teeth, arching its neck, and contracting the lower neck muscles to retract the larynx caudally. This movement is coincided with an in-rush of air through the crico-pharynx into the oesophagus producing the characteristic cribbing sound or grunt. Usually, air is not swallowed but returns to the pharynx. It is considered to be an abnormal, compulsive behavior or stereotypy, and often labelled as a stable vice.

Wind-sucking is a related behavior whereby the horse arches its neck and sucks air into the windpipe but does so without grasping an object. Wind-sucking is thought to form part of the mechanism of cribbing, rather than being defined as an entirely separate behavior.

Wood-chewing

A similar, but unrelated behavior, wood-chewing (lignophagia), is another undesirable behavior sometimes observed in horses. The horse gnaws on wood rails or boards as if they were food, but it does not involve sucking in air.

Prevalence and Incidence

It is reported that 2.4–8.3% of horses in Europe and Canada are cribbers and occupies 15-65% of an individual horse's daily time budget. A postal survey in 2009 found that an average of 4.4% horses in the US are cribbers, but 13.3% of Thoroughbreds perform the behavior. Young Thoroughbred and part-Thoroughbred horses fed concentrated food after weaning are four times more likely to become cribbers than foals not fed concentrate. In several studies, Thoroughbreds consistently have the greatest prevalence of cribbing compared to other breeds. It was found that 11.03% of racehorses performed one or more abnormal stereotypical behaviour that lead back to animal welfare and husbandry systems.

Wind-sucking occurs in 3.8% of non-racing horses in the US. One study shows that stereotypes in general, including cribbing, are more prevalent in dressage horses compared to several other uses.

Geldings and stallions are more likely to exhibit cribbing than mares and the behavior has been reported as occurring in horses on pasture.

Health Effects

There is evidence that stomach ulcers may lead to a horse becoming a cribber, and that cribbing may be a coping mechanism in response to stress.

A 1998 study found that cribbing increased endorphins and found no evidence that cribbing generally impairs the health of affected horses, but later studies reported that cribbing and wind-sucking were related to a history of colic or the subsequent development of colic. A study found that horses would perform the cribbing behaviour in attempt to decrease the cortisol levels that can be brought on by stressful situations. According to this study, the long-term release of stress hormones can be harmful and can cause cardiovascular diseases, depression and immunosuppression.

Causes

Boredom, stress, habit and addiction are all possible causes of cribbing and wind-sucking. It was proposed in a 2002 study that the link between intestinal conditions such as gastric inflammation or colic and abnormal oral behavior was attributable to environmental factors. There is evidence that stomach ulcers may be correlated to a horse becoming a cribber.

Stress is induced when environmental demands produce a physiological response, if that response has a long duration period, it exceeds the normal, natural regulatory ability of the organism. Stress has been found to be a major contributing factor to horses developing this oral stereotypic behaviour. A study suggested that the animal uses cribbing as a coping method when it cannot escape a fearful or stressful situation, or when it has been socially isolated or confined.

Researchers now generally agree that cribbing and wind-sucking occur most often in stabled horses, although once established in an individual horse, the horse may exhibit these behaviors in other places. Recent studies indicate cribbing occurs more frequently in horses that were stable-weaned as foals than in those that were pasture-weaned. In the same study, feeding concentrates after weaning was associated with a fourfold increase in the rate of development of cribbing. The most popular cases of crib-biting come from racetracks, and it is believed to have derived from husbandry systems at the racetracks. The issue with these systems is that the social tendencies of the herd animal have been disrupted. Therefore, they are lacking social interaction and stimulation.

Because thoroughbreds are so consistently the most prevalent cribbers, this suggests there may be a genetic component, however, this may be confounded by different uses and management of different horse breeds. It was found that thoroughbreds are three times more likely to develop this stereotypy than any other breed, supporting that this may be a genetic component. Another study suggesting that cribbing may be genetic found that warmbloods were also more likely to perform this behaviour when compared to other breeds. It was also found that the descendants of a crib-biter were more likely to perform the behaviour due to a genetic component.

Gastrointestinal environment and feeding routines were also a crucial topic, hinting that perhaps grain concentrations, grain ratios and forages were the main cause of ulcers, causing the animal to perform the oral stereotypy as a method of comfort. Horses that were unable to partake in a feeding behaviour that it wants to partake in would influence cribbing. Since the animal is unable to easily digest large quantities of starch, it was found that a high-grain, low forage diet could cause cribbing. A low-forage, high-grain diet was found to increase the risk of the stereotypy because the behaviour aided with relieving stomach acidity.

It has been anecdotally claimed that horses can learn to copy these behaviors from other horses, although this has not been substantiated by scientific study. A study in 2009 found that 48.8% of US horse owners believed that cribbing could be learned by observation, but research demonstrated that only 1.0% of horses developed cribbing after being housed in sight of an affected horse.

Functions

Stereotypies are sometimes considered to be a coping mechanism for animals experiencing stress. A physiological stress response can be induced by injecting an animal with ACTH and the animal's ability to cope with this stress can be monitored by measuring salivary cortisol. In a 2015 study, after ACTH injection, cribbers had higher cortisol levels than non-cribbers. Furthermore, cribbers which did not perform the stereotypy during the 3-hrs of testing had higher cortisol levels than non-cribbers, whereas those performing the stereotypy did not. The researchers concluded that cribbing is a coping mechanism to stressful situations and that because of this, it should not be prevented.

Cribbing and wind-sucking may cause a sensation of pleasure by releasing endorphins in the horse's brain. It has also been suggested that the increase in saliva produced during wind-sucking could be a mechanism for neutralizing stomach conditions in stable-kept, grain-fed horses.

Stereotypies have been defined as "repetitive, invariant behaviour patterns with no obvious goal or function", therefore, if cribbing and wind-sucking have one of the above possible functions, it may be inappropriate to label them as a stereotypy. However, as the causes and resulting reinforcement for these behaviors are probably multifactorial and they remain abnormal behaviors, this indicates that husbandry changes are needed for animals that exhibit cribbing or wind-sucking.

A study suggested was that ghrelin levels were higher in a crib-biting horse than in those who did not perform the behaviour. Cribbing also increased salivary secretion. However, a different study found that there was no relation between salivary secretion and reducing gastric acidity. Meaning that cribbing did not provide comfort for ulcers, rather that it stimulated/caused this issue.

Rearing (Horse)

Rearing occurs when a horse or other equine "stands up" on its hind legs with the forelegs off the ground. Rearing may be linked to fright, aggression, excitement, disobedience, or pain. It is not uncommon to see stallions rearing in the wild when they fight, while striking at their opponent with their front legs. Mares are generally more likely to kick when acting in aggression, but may rear if they need to strike at a threat in front of them.

When a horse rears around people, in most cases, it is considered a dangerous habit for riding horses, as not only can a rider fall off from a substantial height, but also because it is possible for the animal to fall over backwards, which could cause injuries or death to both horse and rider. It is therefore strongly discouraged. A horse that has a habit of rearing generally requires extensive retraining by an experienced horse trainer, and if the habit cannot be corrected, may be deemed too dangerous to ride.

A horse that rears when being handled by a human who is on the ground also presents a hazard, as it is able to strike out with its front feet and can also fall even without the weight of a rider to unbalance the animal. A rearing horse can also break away and escape from a human handler.

However, rearing also has survival value in the wild. It is a tactic that can be used to dislodge a predator that has landed on the animal's back, it is used when equids fight one another, and a horse can rear slightly to add force when striking out with its front feet. For these reasons, horses, particularly young ones, are sometimes seen rearing when loose in a pasture, particularly when playing or mock-fighting with pasturemates.

There are also a few times when rearing is considered acceptable by humans. Rearing may be taught as a trick for circus horses and the like. There are also two movements in classical dressage, the Levade and the Pesade, in which the rider asks the horse to set well back on its hindquarters and raise its front legs off of the ground to varying degrees. However, horses properly trained to do any of these movements are taught to respond to very specific commands and only raise their forequarters when given the command.

Dealing with the Rearing Horse

A horse generally must stop before it can rear. Generally a rider can feel if a horse is about to rear, as the horse shifts its weight strongly to its hindquarters and begins to feel light in the front end. When this occurs, rearing can still be prevented by a number of methods, the simplest of which is to either encourage the horse to move, either forward or to turn the horse in tight circles so that it cannot engage its hindquarters enough to rear. If the horse is allowed to stop or back up while behaving in a disobedient manner, it can more easily rear.

If a horse manages to rear while under saddle, the rider has the best chance of bringing the horse back to the ground by leaning forward, keeping the reins slack and, in some cases, reaching around the neck of the horse to distribute as much weight as possible to the forehand. Once on the ground, the rider can prevent further rearing by asking the horse to move, either forward or in circles.

A horse trained to rear.

A horse (with rider) rearing out of control.

Causes and Solutions

Rearing can be caused by fear; a horse that sees something alarming in front of it may stop and, if asked to continue forward, may rear rather than move ahead. Another fear response may come from poor riding. A rider that is particularly hard on a horse's mouth with the reins and bit may provoke a horse to rear from the pain the horse feels in its mouth. A horse may rear out of confusion because it does not understand what the rider's commands, or riding aids mean, or because the rider is giving harsh or conflicting commands. If a rider both holds onto the horse's mouth at the same time they push the horse strongly with their legs, essentially using the "gas and the brake" at the same time, they can also provoke rearing.

In fact, trained, controlled movements such as the levade and the pesade are deliberately requested by a sophisticated form of collection where a careful, highly balanced rider asks the horse to raise its forequarters by a combination of riding aids that simultaneously gather the horse onto its hindquarters and lighten it in front.

If rearing with a rider is not clearly linked to fear, disobedience or aggression, it may be linked to pain. An equine veterinarian can examine the horse's mouth and teeth, back, and feet for possible causes. Pain may also be linked to poorly fitted or improperly used tack. A rider or saddle-fitting specialist can determine if pain is linked to a poorly fitted saddle, or a broken saddle tree. The fit and severity of the bit can also lead to rearing.

Riders should also consider the management of the horse, especially if the animal does not have turn-out time and has too rich of a diet. A horse may rear due to excitement and excess energy.

For horses that rear while a person is leading them on the ground, the safest position for the handler is to be at the side of the animal so that the handler has maximum control but is still away from the front legs should the horse strike out. Leading horses with a stud chain on the halter or with a bridle offers more control if an animal rears; however, misuse of this equipment by jerking on the horse's head may also provoke rearing.

A rearing horse handled by a person on the ground.

A highly trained horse performing the Pesade, a carefully controlled classical dressage movement where the horse raises its forehand off the ground for a brief period.

Weaving (Horse)

Weaving is a behaviour in horses that is classified as a stable vice, in which the horse repetitively sways on its forelegs, shifting its weight back and forth by moving the head and neck side to side. It may also include swaying of the rest of the body and picking up the front legs. Some horses exhibit non-stereotypical weaving, and instead engage in variations on this behavior.

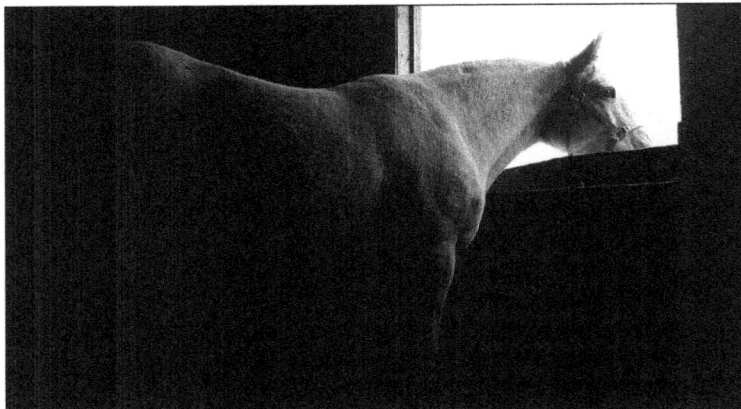

Providing visual stimulation (an open window to the outside) to a stalled horse reduces risk of stable vice occurrence.

Causes

Ultimately, the domestication of horses is considered to be the cause of stable vices such as weaving. There are no reports of wild horses displaying weaving behaviour, mainly because these horses are in their natural state, i.e. they are not confined or on a schedule. Domesticated horses are often housed in stalls (typically 8x8 or 12x12) at night, and are allowed turnout (i.e. time outside) during the day. Horses that are housed in solitary confinement from other horses, or those that do not get daily turnout, or inadequate turnout, are more at risk for developing stable vices such as weaving.

Horses often perform this vice due to stress. Horses tend to weave near the door of their stall, possibly because they desire to leave the stall to move around. Horses also sometimes weave near a window to the aisle or the exterior of the stable, which would provide visual stimulation.

Stress during critical periods such as weaning may also contribute to the development of stables vices.

However, some horses that have developed the habit will also weave while turned out, signifying that the problem is not solely one of confinement.

Many equestrians believe weaving is a learned habit. However, some experts theorize that weaving could more likely develop in horses with a genetic predisposition to it. Thus, there is a debate over whether weaving is a learned behavior that horses pick up by observing another horse who weaves, or if it is an inborn tendency that develops under a certain set of environmental conditions. These two arguments fail to take into account the fact that most behaviours can be both genetically and environmentally influenced. It is possible that both sides are correct to some extent. Horses that exhibit non-stereotypical weaving do not necessarily begin after watching another horse weaving (stereotypical or non-stereotypical), suggesting that horses can begin weaving without learning it from another horse.

Some people claim that it is usually safe to allow other horses to see a weaver, unless it is known that the horse may be genetically predisposed (their sire or dam was a weaver). Others feel it is caused by environmental factors, and that other horses in the same setting will pick up the behavior once a single horse starts. However, this may be due to all horses experiencing similar stresses, and thus engaging in similar behavior.

Weaving may also be caused by anticipation of predictable events such as feeding or turnout, as the behavior has been observed to increase prior to these events.

Negative Effects

Weaving is generally not a very damaging vice over short periods of time, but horses that are consistent weavers may show abnormal hoof wear, and stress on their joints (which can cause lameness). Damage to the stall floor may also occur. The overall value of a horse is not necessarily diminished by its weaving, but the underlying cause of stress or boredom that is causing the behavior should be investigated and rectified to ensure the horse's well-being.

Weaving is also linked to weight loss, uneven muscle development, and performance problems.

Management

Providing a large turnout area, and free-choice hay reduces stress by mimicking a horses natural environment, and therefore reduces risk of weaving behaviour.

Like most vices, weaving is a very difficult habit to break, and may not disappear even after the original problem has been resolved. However, there are several ways to manage a weaver and reduce its stress:

- Allow a weaver to see other horses, even if he is stalled separately.

- Provide a companion for the horse, if possible. Some options include goats, cats, or chickens.

- Provide visual stimulation. In a stall, an open window often helps the situation.

- Keep the horse occupied when stalled. For example, provide a good quality continuous hay or a toy.

- Allow the horse to spend more time outside of its stall. This mimics a horses natural environment and should reduce stress levels.

- Hanging a mirror in a stall often helps weaving, because the horse believes there is a nearby horse. This trick is often very effective, and recent studies in the UK have demonstrated that it can reduce weaving by 97%. Note that the mirror should be made from stainless steel to minimize safety concerns.

- Feed a high quality, high fiber hay and grain to reduce feeding frustration. Consistent feeding times and quality is important, and try to find ways to increase eating time (hay nets).

- Try not to feed wean-age horses concentrated feed (grain), as this can increase stress levels and frustration.

- Alter stall design if horse weaves over door. 'V'-shaped anti-weaving bars prevent weaving. This method is strictly a prevention, and may actually increase the horses frustration.

- Increasing horses exercise, especially if it has limited turnout time during the day.

- Pay particular attention to lowering levels of stress during critical periods in a horses life, especially during weaning. Gradual weaning techniques have been shown to reduce the risk of developing stable vices.

Bucking

Bucking is a movement performed by a horse or bull in which the animal lowers its head and raises its hindquarters into the air, usually while kicking out with the hind legs. If powerful, it may unseat the rider enough to fall off.

Rodeo horse bucking.

Reasons for Bucking

A loose horse may buck due to aggression or fear, as the very high kick of this horse suggests.

Bucking is a normal behavior for a horse with an overabundance of energy,
and in a loose horse, may simply be playful behavior, as here.

Bucking, though a potentially dangerous disobedience when under saddle, is a natural aspect of horse behavior. Bucking developed in the wild for the purpose of protection from feline predators such as mountain lions, who would attack horses by dropping onto their backs from above. The process of kicking out with both hind legs, another defense mechanism for the horse, also results in a mild bucking movement. Thus, for a human to safely ride a horse, the horse has to be desensitized to the presence of something on its back and also learn not to kick out with both hind legs while under saddle. Nonetheless, because the instinct is always there, bucking can still occur for a number of reasons:

- Happiness, such as when a horse bucks during a gallop because of enjoyment, or during play.

- General excitement, such as horses that buck in a crowded schooling ring or at the beginning of a ride in a crowd of horses, such as an endurance ride.

- The rider's aids are causing confusion or fear in the horse, and the horse responds by bucking.

- The horse is fresh, having been kept up in a stall for a long period of time, and is releasing pent-up energy.

- Pain, which may be due to an ill-fitting saddle or another piece of equipment, tooth problems, or other medical issues.

- Provocation, usually due to an insect bite (usually on the hindquarters) which the horse is trying to rid himself of, or in some cases a response to use of a whip on the flank or hindquarters.

- Untrained horses may instinctually buck the first few times they have a saddle on the back if not given proper ground training, and occasionally, even with proper preparation. This is an instinctive defense mechanism.

- Having found that bucking the rider off results in not having to work, the horse does it to avoid his exercise.

- Disobedience to the riding aids, when a horse does not wish to do what is asked by the rider. Sometimes this is due to poor riding on the part of the person, but sometimes a horse attempts to evade a legitimate request by bucking.

- Rodeo broncs are used specifically as bucking horses, usually bred to be prone to bucking and encouraged to buck whenever a rider is on their back with the help of a "bucking strap" around their flank.

- Fear of loud and noisy machines, like cars, trucks, trains, and planes. In response to tragic injuries that have resulted, the American courts have uniformly held that "the needs of a modern, industrial society often conflict with and generally must prevail over the delicate sensibilities of horses."

Ordinary riders need to learn to ride out and correct a simple buck or two, because it is a relatively common form of disobedience. Further, at times, movement akin to bucking is actually required of a horse: Horses that are jumping over an obstacle actually are using almost the same action as bucking when launching themselves into the air, it is simply carried out with advanced planning over a higher and wider distance. The classical dressage movement known as the Capriole is also very similar to the low buck done by a horse when it kicks out with both hind legs.

Solutions to Bucking

Horse bucking as an act of disobedience or discomfort.

Bucking, especially if triggered by fear, pain or excitement, is generally a minor disobedience, unless it is strong enough to unseat the rider, at which point it is a dangerous act. If bucking is a premeditated act of the horse and becomes an undesired habit (such as when a horse learns to buck off a rider so as to no longer have to work), then the horse must be re-schooled by a professional trainer. There have been Olympians who have had to send their horses for re-training by a specialist.

It is important to address the problem of the bucking immediately. Even with good cause, it is a potentially dangerous disobedience that cannot be encouraged or allowed to continue. However, a rider does need to be sure that it is not poor riding that is causing confusion, or a result of poorly-fitting tack that is causing the horse pain. The horse's turn-out schedule should also be assessed, as extra turn-out will give a horse to release his extra energy before the rider gets on. In certain cases (such as a show, when horses are unable to be turned-out for extended periods), longeing the horses for a brief period can help calm him enough so that the rider can get on.

If a horse bucks, the best solution is to use one direct rein to pull the horse's head sideways and up, turning the horse in a small circle. If a rider pulls the horse's head up with both reins, the horse's neck is stronger and the rider is likely to be flipped over the horse's head. By turning the horse sideways, the rider has more leverage and a horse cannot easily buck while turning around. When the horse stops bucking, it must be asked to move forward; a horse also cannot buck very hard while moving forward. Usually a horse gives some warning that it is about to buck by dropping its head, slowing down or stopping, and excessively rounding up in the back (cowboys referred to this as "getting a lump in the back"). To discourage bucking when the rider anticipates it, the rider should ask the horse to move forward or in a circle, raise their hands and the horse's head, and deliberately put the horse into a hollowed-out frame for a moment by sitting back a bit with their heels down, seat deep, and shoulders slightly back. This will help a rider stay in balance if the horse bucks, and the act of deliberately raising the head and hollowing out the horse's back reduces the power and severity of the buck. Certain training aids, such as a gag bit, certain types of martingale or, particularly on ponies, an overcheck, may also discourage bucking.

Consequences of Chronic Bucking

Some horses are chosen for use in rodeos, due to their habitual or powerful bucking ability.

Horses that are chronic and consistent buckers cannot be ridden safely and if they cannot be re-trained become unsuitable for any type of ordinary riding. There are few options available to such an animal, and often humane euthanasia or sale to slaughter is that animal's fate. Horses that cannot be trained not to buck frequently cannot find a home anywhere and may eventually wind up being sold for horsemeat.

In a few cases, a horse that cannot be retrained not to buck may be sold to a rodeo stock contractor. Ironically, such horses often fetch a high price in the bucking stock world because they often are easy to handle on the ground, yet very clever and skilled at unseating riders, thus allowing a cow-boy to obtain a high score if the rider can stay on. At rodeo auctions such as the Miles City Bucking Horse Sale, a spoiled riding horse, particularly one that is powerfully built, will bring a top price and have a long career in rodeo.

Dog Behavior

Dog behavior is the internally coordinated responses of individuals or groups of domestic dogs to internal and external stimuli. It has been shaped by millennia of contact with humans and their lifestyles. As a result of this physical and social evolution, dogs, more than any other species, have acquired the ability to understand and communicate with humans and they are uniquely attuned to their behaviors. Behavioral scientists have uncovered a wide range of social-cognitive abilities in the domestic dog.

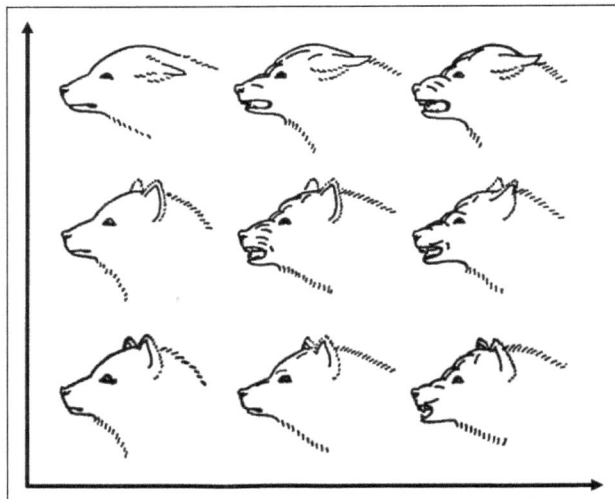

A drawing by Konrad Lorenz showing facial expressions of a dog - a communication behavior.

Co-evolution with Humans

The origin of the domestic dog (*Canis lupus familiaris* or *Canis familiaris*) is not clear. Whole genome sequencing indicates that the dog, the gray wolf and the extinct Taymyr wolf diverged at around the same time 27,000–40,000 years ago. How dogs became domesticated is not clear, however the two main hypotheses are self-domestication or human domestication. There exists evidence of human canine behavioral coevolution.

Intelligence

Dog intelligence is the ability of the dog to perceive information and retain it as knowledge for applying to solve problems. Dogs have been shown to learn by inference. A study with Rico showed that he knew the labels of over 200 different items. He inferred the names of novel items by exclusion learning and correctly retrieved those novel items immediately and also 4 weeks after the initial exposure. Dogs have advanced memory skills. A study documented the learning and memory capabilities of a border collie, "Chaser", who had learned the names and could associate by verbal command over 1,000 words. Dogs are able to read and react appropriately to human body language such as gesturing and pointing, and to understand human voice commands. After undergoing training to solve a simple manipulation task, dogs that are faced with an insolvable version of the same problem look at the human, while socialized wolves do not. Dogs demonstrate a theory of mind by engaging in deception.

Senses

The dog's senses include vision, hearing, sense of smell, taste, touch and sensitivity to the earth's magnetic field.

Communication Behavior

Dog communication is about how dogs speak to each other, how they understand messages that humans send to them, and how humans can translate the ideas that dogs are trying to transmit. These communication behaviors include eye gaze, facial expression, vocalization, body posture (including movements of bodies and limbs) and gustatory communication (scents, pheromones and taste). Humans communicate with dogs by using vocalization, hand signals, and body posture. Dogs can also learn to understand communication of emotions with humans by reading human facial expressions.

Social Behavior

Two studies have indicated that dog behavior varied with their size, body weight and skull size.

Play

Dog-dog

Dog playing with a guinea pig.

Play between dogs usually involves several behaviours that are often seen in aggressive encounters, for example, nipping, biting and growling. It is therefore important for the dogs to place these behaviours in the context of play, rather than aggression. Dogs signal their intent to play with a range of behaviours including a "play-bow", "face-pawed" "open-mouthed play face" and postures inviting the other dog to chase the initiator. Similar signals are given throughout the play bout to maintain the context of the potentially aggressive activities.

From a young age, dogs engage in play with one another. Dog play is made up primarily of mock fights. It is believed that this behavior, which is most common in puppies, is training for important behaviors later in life. Play between puppies is not necessarily a 50:50 symmetry of dominant and submissive roles between the individuals; dogs who engage in greater rates of dominant behaviours (e.g. chasing, forcing partners down) at later ages also initiate play at higher rates. This could imply that winning during play becomes more important as puppies mature.

Emotional contagion is linked to facial mimicry in humans and primates. Facial mimicry is an automatic response that occurs in less than 1 second in which one person involuntary mimics another person's facial expressions, forming empathy. It has also been found in dogs at play, and play sessions lasted longer when there were facial mimicry signals from one dog to another.

Dog-human

The motivation for a dog to play with another dog is distinct from that of a dog playing with a human. Dogs walked together with opportunities to play with one another, play with their owners with the same frequency as dogs being walked alone. Dogs in households with two or more dogs play more often with their owners than dogs in households with a single dog, indicating the motivation to play with other dogs does not substitute for the motivation to play with humans.

It is a common misconception that winning and losing games such as "tug-of-war" and "rough-and-tumble" can influence a dog's dominance relationship with humans. Rather, the way in which dogs play indicates their temperament and relationship with their owner. Dogs that play rough-and-tumble are more amenable and show lower separation anxiety than dogs which play other types of games, and dogs playing tug-of-war and "fetch" are more confident. Dogs which start the majority of games are less amenable and more likely to be aggressive.

Playing with humans can affect the cortisol levels of dogs. In one study, the cortisol responses of police dogs and border guard dogs was assessed after playing with their handlers. The cortisol concentrations of the police dogs increased, whereas the border guard dogs' hormone levels decreased. The researchers noted that during the play sessions, police officers were disciplining their dogs, whereas the border guards were truly playing with them, i.e. this included bonding and affectionate behaviours. They commented that several studies have shown that behaviours associated with control, authority or aggression increase cortisol, whereas play and affiliative behaviour decrease cortisol levels.

Empathy

In 2012, a study found that dogs oriented toward their owner or a stranger more often when the person was pretending to cry than when they were talking or humming. When the stranger pretended to cry, rather than approaching their usual source of comfort, their owner, dogs sniffed, nuzzled and licked the stranger instead. The dogs' pattern of response was behaviorally consistent with an expression of empathic concern.

A study found a third of dogs suffered from anxiety when separated from others.

Personalities

The term personality has been applied to human research, whereas the term temperament has been mostly used for animal research. Personality can be defined by "a set of behaviours that are consistent over context and time". Human personality is often studied using models that look at broad dimensions of personality. For example, the Five Factor Model is one of the most commonly used models, and the most extensively studied. It is composed of five dimensions: openness, conscientiousness, extraversion, agreeableness, and neuroticism. Studies of dog personality have also tried to identify the presence of broad personality traits that are stable over time. Recently, dog's personality dimension has been shown to be relatively consistent over time.

There are different approaches to assess dog personality:

- Ratings of individual dogs: Either a caretaker or a dog expert who is familiar with the dog is asked to answer a questionnaire, for instance the Canine Behavioural Assessment and Research Questionnaire, concerning how often the dog show certain type of behaviour.

- Tests: The dog is submitted to a set of tests and its reactions are evaluated on a behavioural scale. For instance, the dog is presented to a familiar and then an unfamiliar person in order to measure sociability or aggression.

- Observational test: The dog's behaviour is evaluated in a selected but not controlled environment. An observer focus on the dog's reactions to naturally occurring stimuli. For example a walk through the supermarket can allow the observer to see the dog in various types of conditions (crowd, loud noise).

Several potential personality traits have been identified in dogs, for instance "Playfulness", "Curiosity/Fearlessness, "Chase-proneness", "Sociability and Aggressiveness" and "Shyness–Boldness". seven dimension of canine personality:

- Reactivity (approach or avoidance of new objects, increased activity in novel situations).

- Fearfulness (shaking, avoiding novel situations).

- Activity.

- Sociability (initiating friendly interactions with people and other dogs).

- Responsiveness to Training (working with people, learning quickly).

- Submissiveness.

- Aggression.

Dog Breed plays an important role in the dog's personality dimensions, while the effects of age and sex have not been clearly determined. Dogs personality models can be used for a range of tasks, including guide and working dog selection, finding appropriate families to re-home shelter dogs, or selecting breeding stock.

Leadership, Dominance and Social Groups

Two dogs playing follow the leader.

Dominance is a descriptive term for the relationship between pairs of individuals. Among ethologists, dominance has been defined as "an attribute of the pattern of repeated, antagonistic interactions between two individuals, characterized by a consistent outcome in favor of the same dyad member and a default yielding response of its opponent rather than escalation. The status of the consistent winner is dominant and that of the loser subordinate." Another definition is that a dominant animal has "priority of access to resources". Dominance is a relative attribute, not absolute; there is no reason to assume that a high-ranking individual in one group would also become high ranking if moved to another. Nor is there any good evidence that "dominance" is a lifelong character trait. Competitive behavior characterized by confident (e.g. growl, inhibited bite, stand over, stare at, chase, bark at) and submissive (e.g. crouch, avoid, displacement lick/yawn, run away) patterns exchanged.

One test to ascertain in which group the dominant dog was used the following criteria: When a stranger comes to the house, which dog starts to bark first or if they start to bark together, which dog barks more or longer? Which dog licks more often the other dog's mouth? If the dogs get food at the same time and at the same spot, which dog starts to eat first or eats the other dog's food? If the dogs start to fight, which dog usually wins?

Domestic dogs appear to pay little attention to relative size, despite the large weight differences between the largest and smallest individuals; for example, size was not a predictor of the outcome of encounters between dogs meeting while being exercised by their owners nor was size correlated with neutered male dogs. Therefore, many dogs do not appear to pay much attention to the actual fighting ability of their opponent, presumably allowing differences in motivation (how much the dog values the resource) and perceived motivation (what the behavior of the other dog signifies about the likelihood that it will escalate) to play a much greater role.

Two dogs that are contesting possession of a highly valued resource for the first time, if one is in a state of emotional arousal, in pain; if reactivity is influenced by recent endocrine changes, or motivational states such as hunger, then the outcome of the interaction may be different than if none of these factors were present. Equally, the threshold at which aggression is shown may be influenced by a range of medical factors, or, in some cases, precipitated entirely by pathological disorders. Hence, the contextual and physiological factors present when two dogs first encounter each other may profoundly influence the long-term nature of the relationship between those dogs. The complexity of the factors involved in this type of learning means that dogs may develop different expectations about the likely response of another individual for each resource in a range of different situations. Puppies learn early not to challenge an older dog and this respect stays with them into adulthood. When adult animals meet for the first time, they have no expectations of the behavior of the other: they will both, therefore, be initially anxious and vigilant in this encounter (characterized by the tense body posture and sudden movements typically seen when two dogs first meet), until they start to be able to predict the responses of the other individual. The outcome of these early adult–adult interactions will be influenced by the specific factors present at the time of the initial encounters. As well as contextual and physiological factors, the previous experiences of each member of the dyad of other dogs will also influence their behavior.

Scent

Dogs have an olfactory sense 40 times more sensitive than a human's and they commence their lives operating almost exclusively on smell and touch. The special scents that dogs use for communication are called pheromones. Different hormones are secreted when a dog is angry, fearful or confident, and some chemical signatures identify the sex and age of the dog, and if a female is in the estrus cycle, pregnant or recently given birth. Many of the pheromone chemicals can be found dissolved in a dog's urine, and sniffing where another dog has urinated gives the dog a great deal of information about that dog. Male dogs prefer to mark vertical surfaces and having the scent higher allows the air to carry it farther. The height of the marking tells other dogs about the size of the dog, as among canines size is an important factor in dominance.

Dogs (and wolves) mark their territories with urine and their stools. The anal gland of canines give a particular signature to fecal deposits and identifies the marker as well as the place where the dung is left. Dogs are very particular about these landmarks, and engage in what is to humans a meaningless and complex ritual before defecating. Most dogs start with a careful bout of sniffing of a location, perhaps to erect an exact line or boundary between their territory and another dog's territory. This behavior may also involve a small degree of elevation, such as a rock or fallen branch, to aid scent dispersal. Scratching the ground after defecating is a visual sign pointing to the scent marking. The freshness of the scent gives visitors some idea of the current status of a piece of territory and if it is used frequently. Regions under dispute, or used by different animals at different times, may lead to marking battles with every scent marked-over by a new competitor.

Feral Dogs

Feral dogs are those dogs living in a wild state with no food and shelter intentionally provided by humans, and showing a continuous and strong avoidance of direct human contacts. In the developing world pet dogs are uncommon, but feral, village or community dogs are plentiful around humans. The distinction between feral, stray, and free ranging dogs is sometimes a matter of degree, and a dog may shift its status throughout its life. In some unlikely but observed cases, a feral dog that was not born wild but living with a feral group can become rehabilitated to a domestic dog with an owner. A dog can become a stray when it escapes human control, by abandonment or being born to a stray mother. A stray dog can become feral when forced out of the human environment or when co-opted or socially accepted by a nearby feral group. Feralization occurs through the development of a fear response to humans.

Feral dogs are not reproductively self-sustaining, suffer from high rates of juvenile mortality, and depend indirectly on humans for their food, their space, and the supply of co-optable individuals.

Other Behavior

Dogs have a general behavioral trait of strongly preferring novelty (neophillia) compared to familiarity. The average sleep time of a dog in captivity in a 24-hour period is 10.1 hours.

Reproduction Behavior

Estrous Cycle and Mating

Although puppies do not have the urge to procreate, males sometimes engage in sexual play in the form of mounting. In some puppies, this behavior occurs as early as 3 or 4 weeks-of-age.

Dogs reach sexual maturity and can reproduce during their first year, in contrast to wolves at two years-of-age. Female dogs have their first estrus (heat) at 6 to 12 months-of-age; smaller dogs tend to come into heat earlier whereas larger dogs take longer to mature.

Female dogs have an estrous cycle that is nonseasonal and monestrus, i.e. there is only one estrus per estrous cycle. The interval between one estrus and another is, on average, seven months, however, this may range between 4 and 12 months. This interestrous period is not influenced by the photoperiod or pregnancy. The average duration of estrus is 9 days with spontaneous ovulation usually about 3 days after the onset of estrus.

For several days before estrus, a phase called proestrus, the female dog may show greater interest in male dogs and "flirt" with them (proceptive behavior). There is progressive vulval swelling and some bleeding. If males try to mount a female dog during proestrus, she may avoid mating by sitting down or turning round and growling or snapping.

Estrous behavior in the female dog is usually indicated by her standing still with the tail held up, or to the side of the perineum, when the male sniffs the vulva and attempts to mount. This tail position is sometimes called "flagging". The female dog may also turn, presenting the vulva to the male.

The male dog mounts the female and is able to achieve intromission with a non-erect penis, which contains a bone called the os penis. The dog's penis enlarges inside the vagina, thereby preventing its withdrawal; this is sometimes known as the "tie" or "copulatory lock". The male dog rapidly thrust into the female for 1–2 minutes then dismounts with the erect penis still inside the vagina, and turns to stand rear-end to rear-end with the female dog for up to 30 to 40 minutes; the penis is twisted 180 degrees in a lateral plane. During this time, prostatic fluid is ejaculated.

The female dog can bear another litter within 8 months of the previous one. Dogs are polygamous in contrast to wolves that are generally monogamous. Therefore, dogs have no pair bonding and the protection of a single mate, but rather have multiple mates in a year. The consequence is that wolves put a lot of energy into producing a few pups in contrast to dogs that maximize the production of pups. This higher pup production rate enables dogs to maintain or even increase their population with a lower pup survival rate than wolves, and allows dogs a greater capacity than wolves to grow their population after a population crash or when entering a new habitat. It is proposed that these differences are an alternative breeding strategy, one adapted to a life of scavenging instead of hunting.

Parenting and Early Life

All of the wild members of the genus *Canis* display complex coordinated parental behaviors. Wolf pups are cared for primarily by their mother for the first 3 months of their life when she remains in the den with them while they rely on her milk for sustenance and her presence for protection. The father brings her food. Once they leave the den and can chew, the parents and pups from previous years regurgitate food for them. Wolf pups become independent by 5 to 8 months, although they often stay with their parents for years. In contrast, dog pups are cared for by the mother and rely on her for milk and protection but she gets no help from the father nor other dogs. Once pups are weaned around 10 weeks they are independent and receive no further maternal care.

Behavior Problems

There are many different types of behavioural issues that a dog can exhibit, including growling, snapping, barking, and invading human's space. A survey of 203 dog owners in Melbourne, Australia, found that the main behaviour problems reported by owners were overexcitement (63%) and jumping up on people (56%). Some problems are related to attachment while others are neurological, as seen below.

Separation Anxiety

When dogs are separated from humans, usually the owner, they often display behaviors which can be broken into the following four categories: exploratory behaviour, object play, destructive behaviour, and vocalization, and they are related to the canine's level of arousal. These behaviours may manifest as destructiveness, fecal or urinary elimination, hypersalivation or vocalization among other things. Dogs from single-owner homes are approximately 2.5 times more likely to have separation anxiety compared to dogs from multiple-owner homes. Furthermore, sexually intact dogs are only one third as likely to have separation anxiety as neutered dogs. The sex of dogs and whether there is another pet in the home do not have an effect on separation anxiety. It has been estimated that at least 14% of dogs examined at typical veterinary practices in the United States have shown signs of separation anxiety. Dogs that have been diagnosed with profound separation anxiety can be left alone for no more than minutes before they begin to panic and exhibit the behaviors associated with separation anxiety. Separation problems have been found to be linked to the dog's dependency on its owner, not because of disobedience. In the absence of treatment, affected dogs are often relinquished to a humane society or shelter, abandoned, or euthanized.

Resource Guarding

Resource guarding is exhibited by many canines, and is one of the most commonly reported behaviour issues to canine professionals. It is seen when a dog uses specific behaviour patterns so that they can control access to an item, and the patterns are flexible when people are around. If a canine places value on some resource (i.e. food, toys, etc.) they may attempt to guard it from other animals as well as people, which leads to behavioural problems if not treated. The guarding can show in many different ways from rapid ingestion of food to using the body to shield items. It manifests as aggressive behaviour including, but not limited to, growling, barking, or snapping. Some dogs will also resource guard their owners and can become aggressive if the behaviour is allowed to continue. Owners must learn to interpret their dog's body language in order to try to judge the dog's reaction, as visual signals are used (i.e. changes in body posture, facial expression, etc.) to communicate feeling and response. These behaviours are commonly seen in shelter animals, most likely due to insecurities caused by a poor environment. Resource guarding is a concern since it can lead to aggression, but research has found that aggression over guarding can be contained by teaching the dog to drop the item they are guarding.

Noise Anxiety

Canines often fear, and exhibit stress responses to, loud noises. Noise-related anxieties in dogs may be triggered by fireworks, thunderstorms, gunshots, and even loud or sharp bird noises. Associated stimuli may also come to trigger the symptoms of the phobia or anxiety, such as a change in barometric pressure being associated with a thunderstorm, thus causing an anticipatory anxiety.

Tail Chasing

Tail chasing can be classified as a stereotypy. It falls under obsessive compulsive disorder, which is a neuropsychiatric disorder that can present in dogs as canine compulsive disorder. In one clinical study on this potential behavioral problem, 18 tail-chasing terriers were given clomipramine orally at a dosage of 1 to 2 mg/kg (0.5 to 0.9 mg/lb) of body weight, every 12 hours. Three of the dogs required treatment at a slightly higher dosage range to control tail chasing, however, after 1 to 12 weeks of treatment, 9 of 12 dogs were reported to have a 75% or greater reduction in tail chasing. Personality can also play a factor in tail chasing. Dogs who chase their tails have been found to be more shy than those who do not, and some dogs also show a lower level of response during tail chasing bouts.

Behavior Compared to other Canids

Comparisons made within the wolf-like canids allow the identification of those behaviors that may have been inherited from common ancestry and those that may have been the result of domestication or other relatively recent environmental changes. Studies of free-ranging African Basenjis and New Guinea Singing Dogs indicate that their behavioral and ecological traits were the result of environmental selection pressures or selective breeding choices and not the result of artificial selection imposed by humans.

Early Aggression

Dog pups show unrestrained fighting with their siblings from 2 weeks of age, with injury avoided only due to their undeveloped jaw muscles. This fighting gives way to play-chasing with the development of running skills at 4–5 weeks. Wolf pups possess more-developed jaw muscles from 2 weeks of age, when they first show signs of play-fighting with their siblings. Serious fighting occurs during 4–6 weeks of age. Compared to wolf and dog pups, golden jackal pups develop aggression at the age of 4–6 weeks when play-fighting frequently escalates into uninhibited biting intended to harm. This aggression ceases by 10–12 weeks when a hierarchy has formed.

Tameness

Unlike other domestic species which were primarily selected for production-related traits, dogs were initially selected for their behaviors. In 2016, a study found that there were only 11 fixed genes that showed variation between wolves and dogs. These gene variations were unlikely to have been the result of natural evolution, and indicate selection on both morphology and behavior during dog domestication. These genes have been shown to affect the catecholamine synthesis pathway, with the majority of the genes affecting the fight-or-flight response (i.e. selection for tameness), and emotional processing. Dogs generally show reduced fear and aggression compared to wolves. Some of these genes have been associated with aggression in some dog breeds, indicating their importance in both the initial domestication and then later in breed formation.

Social Structure

Among canids, packs are the social units that hunt, rear young and protect a communal territory as a stable group and their members are usually related. Members of the feral dog group are usually not related. Feral dog groups are composed of a stable 2–6 members compared to the 2–15 member wolf pack whose size fluctuates with the availability of prey and reaches a maximum in winter time. The feral dog group consists of monogamous breeding pairs compared to the one breeding pair of the wolf pack. Agonistic behavior does not extend to the individual level and does not support a higher social structure compared to the ritualized agonistic behavior of the wolf pack that upholds its social structure. Feral pups have a very high mortality rate that adds little to the group size, with studies showing that adults are usually killed through accidents with humans, therefore other dogs need to be co-opted from villages to maintain stable group size.

Socialization

The critical period for socialization begins with walking and exploring the environment. Dog and wolf pups both develop the ability to see, hear and smell at 4 weeks of age. Dogs begin to explore the world around them at 4 weeks of age with these senses available to them, while wolves begin to explore at 2 weeks of age when they have the sense of smell but are functionally blind and deaf. The consequences of this is that more things are novel and frightening to wolf pups. The critical period for socialization closes with the avoidance of novelty, when the animal runs away from rather than approaching and exploring novel objects. For dogs this develops between 4 and 8 weeks of age. Wolves reach the end of the critical period after 6 weeks, after which it is not possible to socialize a wolf.

Dog puppies require as little as 90 minutes of contact with humans during their critical period of socialization to form a social attachment. This will not create a highly social pet but a dog that will solicit human attention. Wolves require 24 hours contact a day starting before 3 weeks of age. To create a socialized wolf the pups are removed from the den at 10 days of age, kept in constant human contact until they are 4 weeks old when they begin to bite their sleeping human companions, then spend only their waking hours in the presence of humans. This socialization process continues until age 4 months, when the pups can join other captive wolves but will require daily human contact to remain socialized. Despite this intensive socialization process, a well-socialized wolf will behave differently to a well-socialized dog and will display species-typical hunting and reproductive behaviors, only closer to humans than a wild wolf. These wolves do not generalize their socialization to all humans in the same manner as a socialized dog and they remain more fearful of novelty compared to socialized dogs.

In 1982, a study to observe the differences between dogs and wolves raised in similar conditions took place. The dog puppies preferred larger amounts of sleep at the beginning of their lives, while the wolf puppies were much more active. The dog puppies also preferred the company of humans, rather than their canine foster mother, though the wolf puppies were the exact opposite, spending more time with their foster mother. The dogs also showed a greater interest in the food given to them and paid little attention to their surroundings, while the wolf puppies found their surroundings to be much more intriguing than their food or food bowl. The wolf puppies were observed taking part in antagonistic play at a younger age, while the dog puppies did not display dominant/submissive roles until they were much older. The wolf puppies were rarely seen as being aggressive to each other or towards the other canines. On the other hand, the dog puppies were much more aggressive to each other and other canines, often seen full-on attacking their foster mother or one another.

Cognition

Despite claims that dogs show more human-like social cognition than wolves, several recent studies have demonstrated that if wolves are properly socialized to humans and have the opportunity to interact with humans regularly, then they too can succeed on some human-guided cognitive tasks, in some cases out-performing dogs at an individual level. Similar to dogs, wolves can also follow more complex point types made with body parts other than the human arm and hand (e.g. elbow, knee, foot). Both dogs and wolves have the cognitive capacity for prosocial behavior toward humans; however it is not guaranteed. For canids to perform well on traditional human-guided tasks (e.g. following the human point) both relevant lifetime experiences with humans - including socialization to humans during the critical period for social development - and opportunities to associate human body parts with certain outcomes (such as food being provided by human hands, a human throwing or kicking a ball, etc.) are required.

After undergoing training to solve a simple manipulation task, dogs that are faced with an insoluble version of the same problem look at the human, while socialized wolves do not.

Reproduction

Dogs reach sexual maturity and can reproduce during their first year in contrast to a wolf at two years. The female dog can bear another litter within 8 months of the last one. The canid genus is influenced by the photoperiod and generally reproduces in the springtime. Domestic dogs are not reliant on seasonality for reproduction in contrast to the wolf, coyote, Australian dingo and African basenji that may have only one, seasonal, estrus each year. Feral dogs are influenced by the photoperiod with around half of the breeding females mating in the springtime, which is thought to indicate an ancestral reproductive trait not overcome by domestication, as can be inferred from wolves and cape hunting dogs.

Domestic dogs are polygamous in contrast to wolves that are generally monogamous. Therefore, domestic dogs have no pair bonding and the protection of a single mate, but rather have multiple mates in a year. There is no paternal care in dogs as opposed to wolves where all pack members assist the mother with the pups. The consequence is that wolves put a lot of energy into producing a few pups in contrast to dogs that maximize the production of pups. This higher pup production rate enables dogs to maintain or even increase their population with a lower pup survival rate than wolves, and allows dogs a greater capacity than wolves to grow their population after a population crash or when entering a new habitat. It is proposed that these differences are an alternative breeding strategy adapted to a life of scavenging instead of hunting. In contrast to domestic dogs, feral dogs are monogamous. Domestic dogs tend to have a litter size of 10, wolves 3, and feral dogs 5-8. Feral pups have a very high mortality rate with only 5% surviving at the age of one year, and sometimes the pups are left unattended making them vulnerable to predators. Domestic dogs stand alone among all canids for a total lack of paternal care.

Dogs differ from wolves and most other large canid species as they generally do not regurgitate food for their young, nor the young of other dogs in the same territory. However, this difference was not observed in all domestic dogs. Regurgitating of food by the females for the young, as well as care for the young by the males, has been observed in domestic dogs, dingos and in feral or semi-feral dogs. In one study of a group of free-ranging dogs, for the first 2 weeks immediately after parturition the lactating females were observed to be more aggressive to protect the pups. The male parents were in contact with the litters as 'guard' dogs for the first 6–8 weeks of the litters life. In absence of the mothers, they were observed to prevent the approach of strangers by vocalizations or even by physical attacks. Moreover, one male fed the litter by regurgitation showing the existence of paternal care in some free-roaming dogs.

Space

Space used by feral dogs is not dissimilar from most other canids in that they use defined traditional areas (home ranges) that tend to be defended against intruders, and have core areas where most of their activities are undertaken. Urban domestic dogs have a home range of 2-61 hectares in contrast to a feral dogs home range of 58 square kilometers. Wolf home ranges vary from 78 square kilometers where prey is deer to 2.5 square kilometers at higher latitudes where prey is moose and caribou. Wolves will defend their territory based on prey abundance and pack density, however feral dogs will defend their home ranges all year. Where wolf ranges and feral dog ranges overlap, the feral dogs will site their core areas closer to human settlement.

Predation and Scavenging

Despite claims in the popular press, studies could not find evidence of a single predation on cattle by feral dogs. However, domestic dogs were responsible for the death of 3 calves over one 5-year study. Other studies in Europe and North America indicate only limited success in the consumption of wild boar, deer and other ungulates, however it could not be determined if this was predation or scavenging on carcasses. Studies have observed feral dogs conducting brief, uncoordinated chases of small game with constant barking a technique without success.

In 2004, a study reviewed 5 other studies of feral dogs published between 1975 and 1995 and concluded that their pack structure is very loose and rarely involves any cooperative behavior, either in raising young or in obtaining food. Feral dogs are primarily scavengers, with studies showing that unlike their wild cousins, they are poor ungulate hunters, having little effect on wildlife populations where they are sympatric. However, several garbage dumps located within the feral dog's home range are important for their survival. Even well-fed domestic dogs are prone to scavenge; gastro-intestinal veterinary visits increase during warmer weather as dogs are prone to eat decaying material. Some dogs consume feces, which may contain nutrition. On occasion well-fed dogs have been known to scavenge their owners' corpses.

Dogs in Human Society

Studies using an operant framework have indicated that humans can influence the behavior of dogs through food, petting and voice. Food and 20–30 seconds of petting maintained operant responding in dogs. Some dogs will show a preference for petting once food is readily available, and dogs will remain in proximity to a person providing petting and show no satiation to that stimulus. Petting alone was sufficient to maintain the operant response of military dogs to voice commands, and responses to basic obedience commands in all dogs increased when only vocal praise was provided for correct responses.

A study using dogs that were trained to remain motionless while unsedated and unrestrained in an MRI scanner exhibited caudate activation to a hand signal associated with reward. Further work found that the magnitude of the canine caudate response is similar to that of humans, while the between-subject variability in dogs may be less than humans. In a further study, 5 scents were presented (self, familiar human, strange human, familiar dog, strange dog). While the olfactory bulb/peduncle was activated to a similar degree by all the scents, the caudate was activated maximally to the familiar human. Importantly, the scent of the familiar human was not the handler, meaning that the caudate response differentiated the scent in the absence of the person being present. The caudate activation suggested that not only did the dogs discriminate that scent from the others, they had a positive association with it. Although these signals came from two different people, the humans lived in the same household as the dog and therefore represented the dog's primary social circle. And while dogs should be highly tuned to the smell of items that are not comparable, it seems that the reward response is reserved for their humans.

Research has shown that there are individual differences in the interactions between dogs and their human that have significant effects on dog behavior. In 1997, a study showed that the type of relationship between dog and master, characterized as either companionship or working relationship, significantly affected the dog's performance on a cognitive problem-solving task. They speculate that companion dogs have a more dependent relationship with their owners, and look to them to solve problems. In contrast, working dogs are more independent.

Dogs in the Family

In 2013, a study produced the first evidence under controlled experimental observation for a correlation between the owner's personality and their dog's behaviour.

Dogs at Work

Service dogs are those that are trained to help people with disabilities such as blindness, epilepsy, diabetes and autism. Detection dogs are trained to using their sense of smell to detect substances such as explosives, illegal drugs, wildlife scat, or blood. In science, dogs have helped humans understand about the conditioned reflex. Attack dogs, dogs that have been trained to attack on command, are employed in security, police, and military roles. Service dog programs have been established to help individuals suffering from Post Traumatic Stress Disorder (PTSD) and have shown to have positive results

Attacks

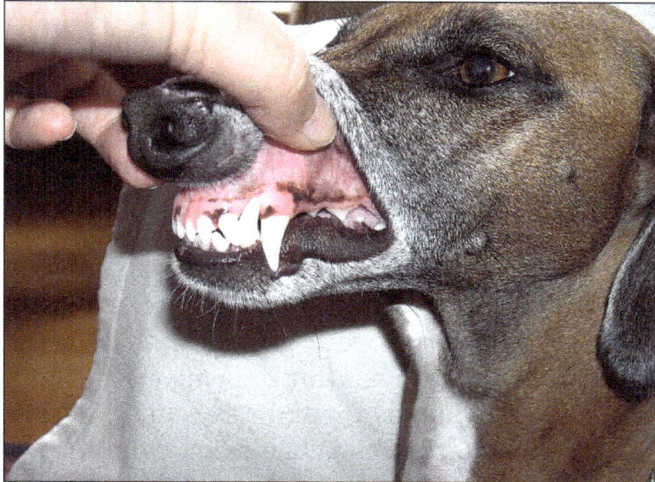

A dog's teeth can inflict serious injuries.

The human-dog relationship is based on unconditional trust; however, if this trust is lost it will be difficult to reinstate.

In the UK between 2005 and 2013, there were 17 fatal dog attacks. In 2007-08, there were 4,611 hospital admissions due to dog attacks, which increased to 5,221 in 2008-09. It has been estimated that more than 200,000 people a year are bitten by dogs in England, with the annual cost to the National Health Service of treating injuries about £3 million. A report published in 2014 stated there were 6,743 hospital admissions specifically caused by dog bites, a 5.8% increase from the 6,372 admissions in the previous 12 months.

In the US between 1979 and 1996, there were more than 300 human dog bite-related fatalities. In the US in 2013, there were 31 dog-bite related deaths. Each year, more than 4.5 million people in the US are bitten by dogs and almost 1 in 5 require medical attention. A dog's thick fur protects it from the bite of another dog, but humans are furless and are not so protected.

Attack training is condemned by some as promoting ferocity in dogs; a 1975 American study showed that 10% of dogs that have bitten a person received attack dog training at some point.

Cat Behavior

Cat behavior includes body language, elimination habits, aggression, play, communication, hunting, grooming, urine marking, and face rubbing. It varies among individuals, colonies, and breeds.

Communication and sociability can vary greatly among individual cats. In a family with many cats, the interactions can change depending on which individuals are present and how restricted the territory and resources are. One or more individuals may become aggressive, fighting may occur with the attack resulting in scratches and deep bite wounds.

A cat's eating pattern in domestic settings are essential for the cat and owner bond to form. This happens because cats form attachments to households that regularly feed them. Some cats ask for food dozens of times a day, including at night, with rubbing, pacing, and meowing, or sometimes loud purring.

Communication

Kittens need vocalization early on in order to develop communication properly. The change in intensity of vocalization will change depending on how loud their feedback is.

Purring or a soft buzz, can mean that the cat is content or possibly that they are sick. Meows are a frequently used greeting. Meows occur when a mother is interacting with her young. Hissing or spitting indicate the cat is angry or defensive. Yowls can mean that the cat is in distress or feeling aggressive. Chattering occurs when they are hunting or being restrained from hunting. If you see your cat making quick chirps, and moving their mouths extremely quickly while their eyes are set and staring at one place, they are chattering, and channeling their inner urge to hunt. Big cats do this as well. Although domesticated cats are not in the wild, they still have their innate need to hunt. Grown cats also do not meow to other grown cats. Cats meow in adult form to talk to other animals, such as dogs, and more importantly humans. Meowing to humans has been researched as that they do it to manipulate humans into what they want and need.

Body Language

Cats greeting by rubbing against each other; the upright "question mark shape" tails also indicate happiness or friendship.

Cats rely strongly on body language to communicate. A cat may rub against an object, lick a person, and purr. Much of a cat's body language is through its tail, ears, head position, and back posture. Cats flick their tails in an oscillating, snake-like motion, or abruptly from side to side, often just before pouncing on an object or animal in what looks like play hunting behavior. If spoken to, a cat may flutter its tail in response, which may be the only indication of the interaction, though movement of its ears or head toward the source of the sound may be a better indication of the cat's awareness that a sound was made in their direction. When cats greet another cat in their vicinity, they can do a slow, languid, long blink to communicate affection if they trust the person or animal they are in contact with. It is a sign of trust. A way to communicate love and trust to your cat from a human perspective is to say their name, and get their attention, and then look them in the eyes and slowly blink at them to emulate trust and love, and they may return the gesture.

Scent Rubbing and Spraying

These behaviors are thought to be a way of marking territory. Facial marking behavior is used to mark their territory as safe. The cat rubs its cheeks on prominent objects in the preferred territory, depositing a chemical pheromone produced in glands in the cheeks. This is known as a contentment pheromone. Synthetic versions of the feline facial pheromone are available commercially.

Cats have anal sacs or scent glands. Scent is deposited on the feces as it is eliminated. Unlike intact male cats, female and neutered male cats usually do not spray urine. Spraying is accomplished by backing up against a vertical surface and spraying a jet of urine on that surface. Unlike a dog's penis, a cat's penis points backward. Males neutered in adulthood may still spray after neutering. Urinating on horizontal surfaces in the home, outside the litter box may indicate dissatisfaction with the box, due to a variety of factors such as substrate texture, cleanliness and privacy. It can also be a sign of urinary tract problems. Male cats on poor diets are susceptible to crystal formation in the urine which can block the urethra and create a medical emergency.

Body Postures

A cat's posture communicates its emotions. It is best to observe cats' natural behavior when they are by themselves, with humans, and with other animals. Their postures can be friendly or aggressive, depending upon the situation. Some of the most basic and familiar cat postures include the following:

- Relaxed posture – The cat is seen lying on the side or sitting. Its breathing is slow to normal, with legs bent, or hind legs laid out or extended. The tail is loosely wrapped, extended, or held up. It also hangs down loosely when the cat is standing.

- Stretching posture – Another posture indicating the cat is relaxed.

Relaxing cat.   Stretching cat.

- Yawning posture – Either by itself, or in conjunction with a stretch: another posture of a relaxed cat.

- Alert posture – The cat is lying on its belly, or it may be sitting. Its back is almost horizontal when standing and moving. Its breathing is normal, with its legs bent or extended (when standing). Its tail is curved back or straight upwards, and there may be twitching while the tail is positioned downwards.

Yawning kitten.

Alert cat.

- Tense posture – The cat is lying on its belly, with the back of its body lower than its upper body (slinking) when standing or moving back. Its legs, including the hind legs are bent, and its front legs are extended when standing. Its tail is close to the body, tensed or curled downwards; there can be twitching when the cat is standing up.

- Anxious/ovulating posture – The cat is lying on its belly. The back of the body is more visibly lower than the front part when the cat is standing or moving. Its breathing may be fast, and its legs are tucked under its body. The tail is close to the body and may be curled forward (or close to the body when standing), with the tip of the tail moving up and down (or side to side).

Tense cat.

Fearful cat.

- Fearful posture – The cat is lying on its belly or crouching directly on top of its paws. Its entire body may be shaking and very near the ground when standing up. Breathing is also fast, with its legs bent near the surface, and its tail curled and very close to its body when standing on all fours.

- Confident posture – The cat may walk around in a more comfortable manner with its tail up to the sky indicating their importance. Cats often walk through houses with their tail standing up high above them making them look grander and more elegant.

- Terrified posture – The cat is crouched directly on top of its paws, with visible shaking seen in some parts of the body. Its tail is close to the body, and it can be standing up, together with its hair at the back. The legs are very stiff or even bent to increase their size. Typically, cats avoid contact when they feel threatened, although they can resort to varying degrees of aggression when they feel cornered, or when escape is impossible.

Terrified cat.

Grooming

Oral grooming for domestic and feral cats is a common behavior; recent studies on domestic cats show that they spend about 8% of resting time grooming themselves. Grooming is extremely important not only to clean themselves but also to ensure ectoparasite control. Fleas tend to be the most common ectoparasite of cats and some studies allude to indirect evidence that grooming in cats is effective in removing fleas. Cats not only use their tongue for grooming to control ectoparasites but scratch grooming as well may aid in dislodging fleas from the head and neck.

Kneading

Kittens "knead" the breast while suckling, using the forelimbs one at a time in an alternating pattern to push against the mammary glands to stimulate lactation in the mother.

Cats carry these infantile behaviors beyond nursing and into adulthood. Some cats "nurse", i.e. suck, on clothing or bedding during kneading. The cat exerts firm downwards pressure with its paw, opening its toes to expose its claws, then closes its claws as it lifts its paw. The process takes place with alternate paws at intervals of one to two seconds. They may knead while sitting on their owner's lap, which may prove painful if the cat has sharp claws.

Since most of the preferred "domestic traits" are neotenous or juvenile traits that persist in the adult, kneading may be a relic juvenile behavior retained in adult domestic cats. It may also stimulate the cat and make it feel good, in the same manner as a human stretching. Kneading is often a precursor to sleeping. Many cats purr while kneading. They also purr mostly when newborn, when feeding, or when trying to feed on their mother's teat. The common association between the two behaviors may corroborate the evidence in favor of the origin of kneading as a remnant instinct.

Panting

Unlike dogs, panting is a rare occurrence in cats, except in warm weather environments. Some cats may pant in response to anxiety, fear or excitement. It can also be caused by play, exercise, or stress from things like car rides. However, if panting is excessive or the cat appears in distress, it may be a symptom of a more serious condition, such as a nasal blockage, heartworm disease, head trauma, or drug poisoning. In many cases, feline panting, especially if accompanied by other symptoms, such as coughing or shallow breathing (dyspnea), is considered to be abnormal, and treated as a medical emergency.

A cat panting.

Reflexes

Righting Reflex

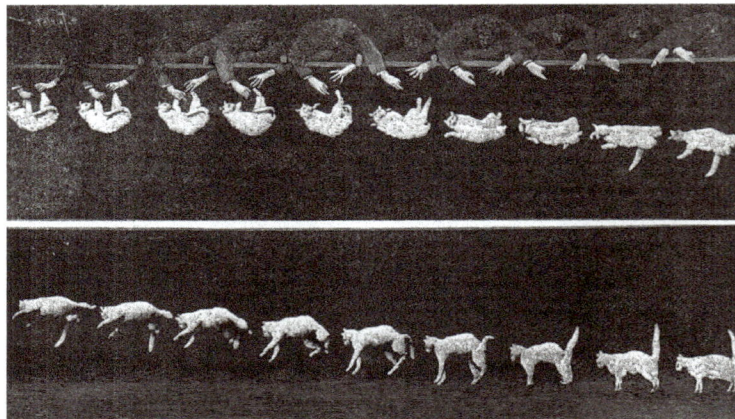

Chronophotography of a falling cat.

The righting reflex is the attempt of cats to land on their feet at the completion of a jump or a fall. They can do this more easily than other animals due to their flexible spine, floating collar bone, and loose skin. Cats also use vision and their vestibular apparatus to help tell which way to turn. They can then stretch themselves out and relax their muscles. The righting reflex does not always result in the cat landing on its feet at the completion of the fall.

Freeze Reflex

Adult cats are able to make use of pinch-induced behavioural inhibition to induce a 'freeze reflex' in their young which enables them to be transported by the neck without resisting. This reflex can also be exhibited by adults. This is also known as 'clipnosis'.

Eating Patterns

Cat eating "cat grass".

Cats are obligate carnivores, and do not do well on vegetarian diets. In the wild they usually hunt smaller mammals to keep themselves nourished. Many cats find and chew small quantities of long grass but this is not for its nutritional value per se. The eating of grass seems to stem from feline ancestry and has nothing to do with dietary requirements. It is believed that feline ancestors instead ate grass for the purging of intestinal parasites.

Cats have no sweet taste receptors on their tongue and thus cannot taste sweet things at all. Cats mainly smell for their food and what they taste for is amino acids instead. This may be a cause of cats being diagnosed with diabetes. The food that domestic cats get has a lot of carbohydrates in it and a high sugar content cannot be efficiently processed by the digestive system of cats.

Cats drink water by lapping the surface with their tongue. A fraction of a teaspoon of water is taken up with each lap. Although some desert cats are able to obtain much of their water needs through the flesh of their prey, most cats come to bodies of water to drink.

Eating patterns is another indicator to understand the behavior changers in domestic cats. The changers in typical eating patterns can be a early signal for possible physical or psychological health problem.

Excretion

Cats tend to bury their feces after defecating and can be attracted to a litter box if it has attractant in it. Cats will generally defecate more in those litter boxes too.

Socialization

Socialization is defined as a member of a specific group learning to be part of that said group. It is said to be a continuous learning process that allows an individual to learn the necessary skills and behaviours required for a particular social position.

Cats, domestic or wild, do participate in social behaviours, even though it is thought that most cat species (besides lions) are solitary, anti-social animals. Under certain circumstances, such as food availability, shelter, or protection, cats can be seen in groups.

The social behaviours that cats participate in are colony organization, social learning, socialization between cats, and socialization with humans.

Colony Organization

Free-living domestic cats tend to form small to large colonies. Small colonies consist of one female, known as a queen, and her kittens. Large colonies consist of several queens and their kittens. Male cats are present in both types of colonies and serve the purpose of reproduction and defending territory. Within these colonies altruistic behaviour occurs. This means that if an expecting queen helps another queen that just gave birth, then the helping queen will get help when she gives birth in return.

Although free living cats are found in colonies, stable social order, like that of the lion, does not exist. Free living cats usually are found in colonies for protection against predators, and food availability. Although there are many advantages of group living, such as easy access to mates, and defensive measures to protect food, there are also disadvantages, such as sexual competition for mates, and if the group becomes too big then fights may break out over food.

Social Learning

Cats are observational learners. This type of learning emerges early in a cat's life, and has been shown in many laboratory studies. Young kittens learn to hunt from their mothers by observing their techniques when catching prey. The mother ensures their kittens learn hunting techniques by first bringing dead prey to the litter, then live prey. With the live prey, she demonstrates the techniques required for successful capture to her kittens by bringing the live prey to the litter for the kittens to catch themselves. Prey-catching behaviour of kittens improves over time when mothers are present over when they are not.

Observational learning for cats can be described in terms of the drive to complete the behaviour, the cue that initiates the behaviour, the response to the cue, and the reward for completing the behaviour. This is shown above when cats learn predatory behaviour from their mothers. The drive is hunger, the cue is the prey, the response is to catch the prey, and the reward is to relieve the hunger sensation.

Kittens also show observational learning when they are socializing with humans. They are more likely to initiate socialization with humans when their mothers are exhibiting non-aggressive and non-defensive behaviours. Even though mothers spend most time with their kittens, male cats play an important role by breaking up fights among litter mates.

Observational learning is not limited to kitten-hood, it can also be observed during adulthood. Studies have been done with adult cats performing a task, such as pressing a lever after a visual cue. Adult cats that see others performing a task learn to perform the same task faster than those who did not witness another cat.

Socialization between Cats

When strange cats meet, ideally they would cautiously allow the strange cat to smell its hindquarters, but this does not happen very often. Usually when strange cats meet, one cat makes a sudden movement that puts the other cat into a defensive mode. The cat will then draw in on itself and prepare to attack if needed. If an attack were to happen the subordinate cat will usually run away, but this does not happen all the time and it could lead to a tomcat duel. Dominance is also seen as an underlying factor for how conspecifics interact with each other.

Socialization and Communication between Cats and Humans

Cats have learned how to develop their vocals in order to converse with humans, in which they try to tell humans what they want. Cats vocalize to other cats. Hisses and spits warn other cats to keep their distance. It may be noted that a "hiss" is less about the sound and more about the showing of teeth along with their stance. Kittens also meow at their mothers for milk and attention, but this may go away once they have no need for milk as they become adults if they are not actively socialized. Another way that cats and humans interact is through what people call "head bunting" in which a cat rubs its head on a human in order to leave their scent, mark to claim territory, and create a bond.

Dominance

Dominance can be seen among cats in multi-cat households. It can be seen when other cats submit to the "dominant" cat. Dominance includes such behaviours as walking around the dominant cat, waiting for the dominant cat to walk past, avoiding eye contact, crouching, laying on their side (defensive posture), and retreating when the dominant cat approaches. Dominant cats present a specific body posture as well. The cat displays ears straight up, the base of its tail will be arched, and it looks directly at subordinate cats. These dominant cats are usually not aggressive, but if a subordinate cat blocks food they may become aggressive. When this aggressive behaviour occurs it could also lead to the dominant cat preventing subordinate cats from eating and using the litter box. This can cause the subordinate cat to defecate somewhere else and create problems with human interaction.

Social Conflicts

Two cats fighting.

Social conflicts among cats depend solely on the behaviour of the cats. Some research has shown that cats rarely pick fights, but when they do its usually for protecting food and litters, and defending territory.

The first sign of an imminent tomcat duel is when both cats draw themselves up high on their legs, all hair along the middle of their backs is standing straight up, and they mew and howl loudly as they approach one another. The steps the cats make become slower and shorter the closer they become to one another. Once they are close enough to attack, they pause slightly, and then one cat leaps and tries to bite the nape of the other cat. The other cat has no choice but to retaliate and both cats roll aggressively on the ground, and loud intense screams come from both cats. After some time the cats separate and stand face to face to begin the attack all over again. This can go on for some time until one cat does not get up again and remains seated. The defeated cat does not move until the victor has completed a sniff of the area and moves outside the fighting area. Once this happens the defeated cat leaves the area, ending the duel.

Females may also fight with each other. If a male and female do not get along, they may also fight. Cats may need to be reintroduced or separated to avoid fights in a closed household.

Socialization with Humans

Cats between the age of three to nine weeks are sensitive to human socialization; after this period socialization can be less effective. Studies have shown that the earlier the kitten is handled, the less fearful the kitten will be towards humans. Other factors that can enhance socialization are having many people handle the kitten frequently, the presence of the mother, and feeding. The presence of the mother is important because cats are observational learners. If the mother is comfortable around humans then it can reduce anxiety in the kitten and promote the kitten-human relationship.

Feral kittens around two to seven weeks old can be socialized usually within a month of capture. Some species of cats cannot be socialized towards humans because of factors like genetic influence and in some cases specific learning experiences. The best way to get a kitten to socialize is to handle the kitten for many hours a week. The process is made easier if there is another socialized cat present but not necessarily in the same space as the feral. If the handler can get a cat to urinate in the litter tray, then the others in a litter will usually follow. Initial contact with thick gloves is highly recommended until trust is established, usually within the first week. It is a challenge to socialize an adult. Socialized adult feral cats tend to trust only those who they trusted in their socialization period, and therefore can be very fearful around strangers.

Cats are also used for companion animals. Studies have shown that these animals provide many physiological and psychological benefits for the owner. Other aspects of cat behaviour that are deemed advantageous for the human–cat bond are cat hygiene (cats are known for good hygiene), they do not have to be taken outside (use of the litter box), they are perfect for smaller spaces, and they have no problems with being left alone for extended periods of time. Even though there are a number of benefits for owning a cat, there are a number of problematic behaviours that affect the human–cat relationship. One behaviour is when cats attack people by clawing and biting. This often occurs spontaneously or could be triggered by sudden movements. Another problematic behaviour is the "petting and biting syndrome", which involves the cat being pet and then suddenly attacking and running away. Other problems are house soiling, scratching furniture, and when a cat brings dead prey into the house. It is these kinds of behaviours that put a strain on the socialization between cats and people.

There are 52 measured cat personality traits in cats, with one study suggesting "five reliable personality factors were found using principal axis factor analysis: neuroticism, extraversion, dominance, impulsiveness and agreeableness."

Predatory Behavior

Cats are natural predators. When allowed to roam outdoors, many cats will engage in predation on wildlife. When kept as an indoor pet due to high-density living, traffic, or safety from predators such as coyotes, they are essentially captives, like zoo animals. Understanding an indoor cat's personality can go a long way to satisfy their instincts and avoid potentially inconvenient behavior (such as sudden hissing, dashing around the house, or climbing the curtains). While there may not be common rules for providing a stable environment it appears that the following should be present:

- A good-sized cat tree, with scratching posts.

- Toys that provide a release for their predatory instincts (a length of string dragged around is very popular, although string is a dangerous toy for cats).

- A well kept litter box or toilet.

- Fresh water and dry cat food.

- Social interaction.

Environment

Cats like to organize their environment based on their needs. Like their ancestors, domestic cats still have an inherent desire to maintain an independent territory but are generally content to live with other cats for company as they easily get bored. Living alone for a longer time may cause them to forget how to communicate with other cats.

Sometimes, however, adding a kitten to a household can be a bad idea. If there already is an older cat present and another cat is added to their environment it may be better to get another older cat that has been socialized with other cats. When a kitten is introduced to a mature cat, that cat may show feline asocial aggression where they feel threatened and act aggressive to drive off the intruders. If this happens, the kitten and the cat should be separated, and slowly introduced by rubbing towels on the animals and presenting the towel to the other.

Cats use scent and pheromones to help organize their territory by marking prominent objects. If these objects or scents are removed it upsets the cat's perception of its environment.

Importance of Space for Domestic and Feral Cats

The domestic cat has become more sociable through contact of its own species through domestication. The domestic cat is more juvenile than the African wildcat; this promotes greater tolerance of other cats and domestic animals.

It has been documented that feral cat colonies have a social structure where the females of the group live together and help with each other's kittens whereas the males do not. There are also studies that cats do form hierarchies when housed in a limited space. It is known that cats show higher levels of stress during the first couple of weeks at a shelter vs. in a group house when controlled together for 2–16 weeks. those cats housed in smaller spaces are forced to interact with each other while the more space per cat the less stress-related behaviours. The experiment approved by the Swedish Ethical Committee in Gothenburg, concluded. From our result we conclude that increasing the space for group-housed cats, from 1 m²/cat to 4 m²/cat, increases the amount of play behavior. The amount of licking and body contact (i.e. positive inactivity) between cats, and the activity increases when cats are housed on 4 m²/cat compared to 2 m²/cat. Play has been used as an indicator of positive welfare in other species, and licking and body contact can indicate positive contact between individuals, we therefore argue that the increase in space given to these cats slightly increased their welfare. However, further studies are needed on the effect of space and density of cats when housed together in groups.

Cat Scratching

It may seem like kitty is scratching your couch and curtains to annoy you, but she's really doing it to work off energy, to play, to mark her territory, even to get rid of frayed bits of claw. Good news: "Scratching is easy to prevent,". So you don't have to settle for raggedy furniture or stop kitty from expressing her natural behavior. To prevent scratching damage:

- Buy one or more scratching posts for your cat, then dab a bit of catnip on the posts to entice your feline friend to use them.

- Trim your kitty's claws. It may seem daunting, but trimming is easier than you think. Get a quick tutorial from your veterinarian, who can probably do the deed in 10 seconds a skill that can be learned.

- Turn your cat into a fashion plate with colorful claw caps (also called nail caps). These small, vinyl sleeves fit over kitty's claws, preventing them from doing damage when they scratch.

Cat tearing up your favorite loveseat? You might want to invest in a nice scratching post.

Cats scratch furniture for a number of reasons. Scratching is a good form of exercise for them. They get to stretch out their bodies and extend and retract their nails. When they scratch, the movements help remove the outer nail sheaths. Cats also scratch to leave visual and olfactory (scent) markers. Their interdigital glands, which are located between the pads of their paws, leave odors behind so that other cats know that the marker cat has been in the area. When cats scratch objects, they also leave small gouges, which are visual signals to other cats that there is a cat in the area.

Some cats also may scratch furniture because they are not provided with adequate scratching posts. Other cats have developed a preference for particular materials, such as the expensive fabrics that our couches and armchairs are made from! The location of certain furniture may make it a great place for cats to provide a visual signal or get a good scratch in after waking from a nap. In the wild, scratching sites are usually located in areas where the cats spend a fair portion of their time.

References

- Mintz, Zoe (14 January 2014). "Humans And Primates Burn 50 Percent Fewer Calories Each Day Than Other Mammals". Www.ibtimes.com. IBT Media Inc. Retrieved 2014-01-14

- Reproduction, mammal, animal: britannica.com, Retrieved 6 January, 2019

- Janečka, J. E.; Miller, W.; Pringle, T. H.; Wiens, F.; Zitzmann, A.; Helgen, K. M.; Springer, M. S.; Murphy, W. J. (2 November 2007). "Molecular and Genomic Data Identify the Closest Living Relative of Primates". Science. 318 (5851): 792–794. Bibcode:2007Sci...318..792J. Doi:10.1126/science.1147555. PMID 17975064

- Primate-social-organization, chapter, the-history-of-our-tribe-hominini: milnepublishing.geneseo.edu, Retrieved 26 February, 2019

- Groves, C.P. (2005). Wilson, D.E.; Reeder, D.M. (eds.). Mammal Species of the World: A Taxonomic and Geographic Reference (3rd ed.). Baltimore: Johns Hopkins University Press. Pp. 111–184. ISBN 0-801-88221-4. OCLC 62265494

- Common-cat-problems-how-solve-them, features, cats: pets.webmd.com, Retrieved 7 March, 2019

- Breuer, T.; Ndoundou-Hockemba, M.; Fishlock, V. (2005). "First observation of tool use in wild gorillas". PLOS Biology. 3 (11): e380. Doi:10.1371/journal.pbio.0030380. PMC 1236726. PMID 16187795

- Why-does-my-cat-scratch-furniture, our-pet-experts: vetstreet.com, Retrieved 8 April, 2019

- Marchant, Linda Frances; Nishida, Toshisada (1996). Great ape societies. Cambridge University Press. Pp. 226–227. ISBN 978-0-521-55536-4. Retrieved 4 July 2011

- E. Fish, Frank (2002). "Balancing Requirements for Stability and Maneuverability in Cetaceans". Integrative and Comparative Biology. 42 (1): 85–93. Doi:10.1093/icb/42.1.85. PMID 21708697

Diseases of Mammals | 5

- **Zoonotic Diseases in Mammals**
- **Primate Diseases**
- **Horse Diseases**
- **Dog Diseases**
- **Cat Diseases**

There are various diseases which are caused in different mammals like horses, dogs, cats, etc. Potomac horse fever, skin cancer, infectious canine hepatitis, hypertrophic osteodystrophy, congenital sensorineural deafness, feline asthma, etc. are some of these diseases. The topics elaborated in this chapter will help in gaining a better perspective about these diseases caused in mammals.

Zoonotic Diseases in Mammals

Across the 27 terrestrial mammal orders, there is wide variation in both the total number of species, and the fraction of species in each order that are zoonotic hosts. In the most speciose orders, the number of zoonotic hosts increases with total species richness of the order, a pattern which may be explained in part by accumulating evidence that parasitism drives host diversification. But further examination of the variation around this general trend is warranted. For example, groups with more zoonotic host species than expected for the richness of the clade (e.g., orders that fall above a regression line through points in figure) may share suites of similar intrinsic or extrinsic factors enabling more successful pathogen transmission, or more frequent human contacts that, over time, facilitate the transition of a novel infection to a disease state in humans (e.g., the ungulates, a paraphyletic group that includes the Artiodactyla and Perissodactyla; ungulates comprise the majority of domesticated mammal species). Conversely, orders with fewer zoonotic hosts than expected for the richness of the clade (those below a regression line) may reflect species that have been poorly sampled for zoonoses, or species that carry fewer zoonoses due to unique combinations of intrinsic and extrinsic features. For example, the marsupial carnivores (Dasyuromorpha), much like placental carnivores (Carnivora), have less direct contact with humans compared to other clades (e.g., Rodents, Ungulates), which may reduce the risk of zoonotic transmission to humans. However, many dasyuromorphs have regular contact with domesticated species (for example, domesticated dogs, especially when Tasmanian devils and quolls are attracted to livestock on farms), providing opportunity for human exposure through farm animals on human-modified environments.

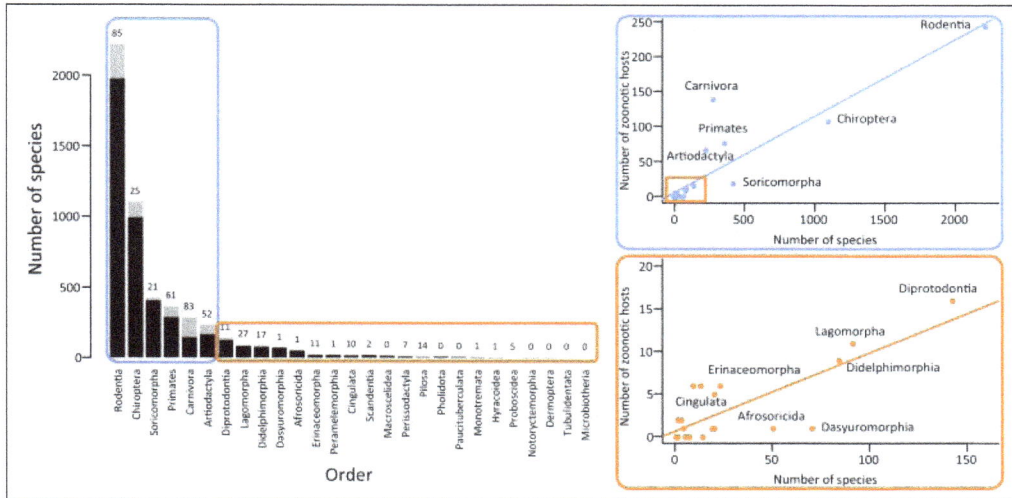

The number of zoonotic hosts increases with total species richness of the order.

Zoonotic diseases are found in the majority of terrestrial mammal orders (21/27), with the most species rich orders containing the greatest diversity of zoonoses. This split bar plot shows the total number of host species (black and gray) and the fraction of species that are confirmed zoonotic hosts for one or more zoonotic diseases (gray). The number above each bar represents a tally of the total unique zoonoses per order. Mammal orders are arranged in descending order of species richness. The number of zoonotic host species in each order is represented in scatterplots, with the most speciose orders contained in the blue boxes (top right; regression $R^2=0.81$), and all other orders in the orange boxes (bottom right; regression $R^2=0.63$).

Maps depicting geographic range distributions of zoonotic hosts across the most speciose orders show that global hotspots are driven in part by striking differences in the distribution of zoonotic hosts from specific clades, as will be explained below. Maps depicting geographic range distributions of zoonotic hosts from mammal orders endemic to Australia can be found in figure, and maps for all other orders can be found in figure.

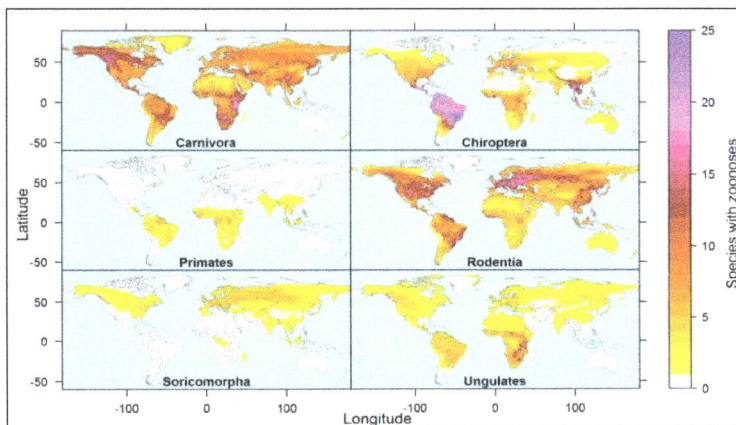

Global hotspots of zoonoses are driven by differences in the distribution of zoonotic hosts from specific clades.

Mapping overlapping geographic ranges of mammal species recognized to carry one or more zoonotic diseases highlights regions of high and low zoonotic host diversity arising from particular clades. Mammal zoonotic host richness is depicted by color for carnivores, bats (Chiroptera), primates, rodents, shrews and moles (Soricomorpha), and the hoofed mammals (Ungulates, which combine the orders Perissodactyla and Artiodactyla and exclude domesticated species).

Rodentia

Among mammals, rodents are the most abundant and most species-rich and include a greater number of zoonotic hosts than any other order: approximately 10.7% of rodents are hosts (244/2220 species, updated from proportions reported by), carrying 85 unique zoonotic diseases. Rodent reservoirs of zoonotic diseases are distinguished by features that support a fast life history profile, reproducing earlier in life and more frequently compared to other rodent species. Figure, shows that north temperate areas in North America and Europe and the tropical Atlantic forest region of Brazil contain the most rodent host species. We note that the larger global zoonotic host hotspot observed in Europe and Russia may be driven in part by the diversity of rodent and small-bodied insectivore hosts (Soricomorpha, see below), as well as their predators.

Chiroptera

Because they are also relatively small-bodied, speciose, and associated with numerous EID events in humans, bats are often compared to rodents with respect to their risk of carrying zoonotic pathogens. There are about half as many zoonotic bat hosts compared to rodents (108/1100 bat species are hosts, approximately 9.8%), and they carry about a third of the number of unique zoonoses compared to rodents.

The mammal host hotspot in the Neotropics is likely to be driven in part by the high diversity of bat hosts in this region. Figure, shows hotspots of zoonotic bat hosts in Central and South America (wet regions east of Chile, north of Paraguay and Uruguay), as well as in Southeast Asia. These patterns are generally consistent with bat biodiversity patterns, except for the following departures. 1) within South America, bat species richness is highest in the Andean countries and in northern Brazil whereas the hotspot of bat hosts is in southern Brazil; 2) Southeast Asia and equatorial Africa display similar patterns of bat species richness, but Southeast Asia has many more zoonotic host species even though it is a much smaller land mass. If, as ecological theory suggests, we expect the number of hosts to be proportional to overall species richness, these maps suggest either that Africa is understudied or that Southeast Asia has more zoonotic hosts than expected for its mammal species richness and for its area.

Soricomorpha

Among the insectivoran mammals, relatively few shrews and moles are known to be zoonotic hosts, with only about 4% (19/426 species) carrying 19 unique zoonoses. This small percentage could be due to this group being understudied compared to other, similarly species rich mammal groups. A Web of Science search on the Latin binomials of all extant mole and shrew species returned a total of 4600 citations, which is 1-2 orders of magnitude fewer studies than for other speciose mammal orders. Zoonotic hosts in this order are distributed widely across north temperate latitudes, with the greatest number of host species overlapping across Europe and across the Atlantic coast of the United States.

Carnivora

While ungulates were previously thought to share the most pathogens with humans, we find that carnivoran hosts nearly tie with the rodents to harbor more unique zoonoses than other terrestrial mammal clades. Approximately 49% (139/285) of all carnivore species the highest proportion of any mammal order carry one or more of 83 unique zoonotic pathogens. Carnivoran hosts are among the most widely distributed in terms of spatial extent, with hotspots of host diversity in Southern and East Africa, Southeast Asia, and the subarctic region of North America. This contrasts patterns of carnivoran species richness, which is concentrated in the southern hemisphere. pathogen richness (zoonotic and non-zoonotic) closely tracked carnivore species richness, and that the range of carnivoran host species infected by a pathogen depends primarily on host phylogenetic relatedness.

Ungulates

Ungulate reservoirs of zoonotic disease have been of particular interest because of high human contact rates through hunting, and the degree of contact and relatedness between wild and domesticated species (livestock). Recent work also shows that the time since domestication correlates positively with the number of zoonotic infections shared between ungulates and humans, and that species with the longest history of domestication do not only carry more zoonotic pathogens, but may also transmit infection to a greater diversity of alternative host species. For wild ungulates (excluding domesticated species), we find that approximately 32% of species were zoonotic hosts (73/247 species), carrying 68 unique zoonoses. Ungulates cover a greater spatial extent than bats, primates and insectivores, and the majority of host species overlap in East and Southern Africa.

Primates

The high degree of phylogenetic relatedness between human and non-human primates is thought to contribute to greater risk of pathogen spillover. For example, species that are closely related and share habitat show the most similar parasite communities, suggesting that spatial overlap and phylogenetic relatedness are likely to be important for understanding transmission in humans and in wild host species. Primates are generally found in the global equatorial zone, with greatest species diversity in the rainforests of Africa, the Neotropics, and Asia. Primate zoonotic host richness is greatest in equatorial Africa (in central Africa in the Congo Basin, and West Africa), in Southeast Asia, and in the tropical/mixed forest regions of northern Brazil and the Guyana Shield. Over 20% of primate species are zoonotic hosts (21%; 77/365 species) for at least one of 63 unique zoonoses. Thus, while there are fewer species of primates overall, a greater proportion of primate are zoonotic hosts than either the rodents or the bats.

Global Hotspots of Mammal-borne Zoonoses

If we make the simplifying assumption that zoonoses are distributed throughout the geographic ranges of the mammals carrying them, then we can generate some baseline hypotheses about where zoonotic potential may be greatest by identifying zoonotic pathogen hotspots. Such hotspots occur where many zoonotic hosts overlap in geographic range, and thus their zoonotic pathogens also overlap. The histograms in figure show that species richness of zoonotic hosts reflects latitudinal gradients reminiscent of well-known biogeographical patterns in free-living organisms (more host species at lower latitudes). However, the richness of zoonotic diseases does not exhibit this pattern: despite wide variation in the global distribution of land mass and species richness, chi-squared tests showed that the numbers of unique zoonoses found across 30-degree longitudinal bands were not significantly different from each other ($\chi 211=17.571$, p-value=0.09; similar analyses across 30-degree bands of latitude and longitude showed expected geographical patterns of higher richness of hosts and zoonoses in the tropics, and in the longitudinal bands capturing greater land mass, table). Compared to zoonotic mammal hosts, zoonoses are distributed more evenly worldwide suggesting that, compared to wildlife-specific pathogens whose ranges are delimited by ecological interactions regulating their host and vector populations, the distributions of zoonoses are not as tightly coupled to the distribution of their host species. Zoonoses are not as broadly distributed as human-specific diseases, but are possibly more labile than wildlife diseases due to the mobility and range expansion of human populations while adhering to biogeographic grouping patterns reflecting ecological barriers to animal host species establishment and dispersal.

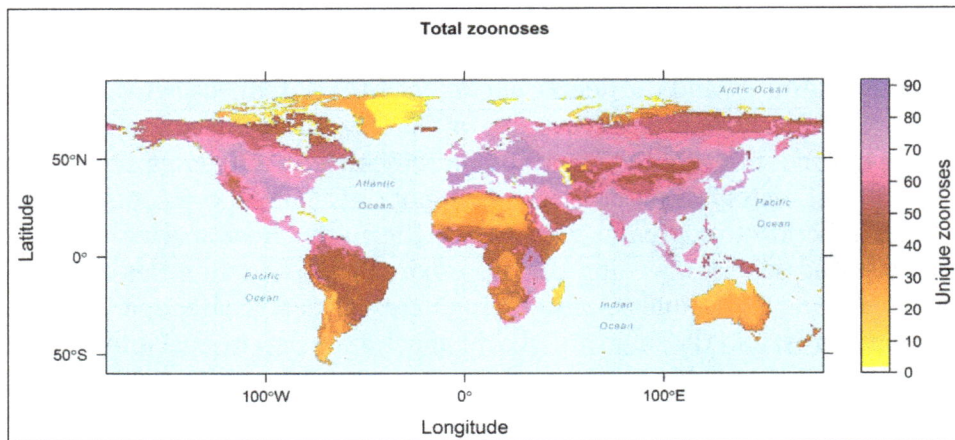

Overlapping geographic ranges of zoonotic diseases carried by wild terrestrial mammal host species from 27 orders.

Though major hotspots of mammal hosts occur in the New and Old World tropics (South America and Eastern Africa, particularly; figure), there are more zoonoses concentrated in northern latitudes, Eastern Africa, and Southeast Asia. This is opposite to the patterns depicted for zoonotic hosts in the tropics, where host richness is expected to match global patterns of high species richness that increase the frequency of consumer-resource interactions overall, including parasitic interactions. In addition to the hotspots of human emerging diseases observed in the tropics and Europe, figure draws attention to the global subarctic. While this region has lower zoonotic host and species diversity compared to other biogeographic regions, mammals found in the subarctic zone harbor more zoonoses than hosts from other regions. In general, mammal species are predominantly constrained by their abiotic environments, but pathogens contend primarily with the biotic environment presented by their hosts. Thus, one possible explanation of this pattern is that while there are fewer host species in the global subarctic compared to other regions of the world, the pathogens causing zoonoses in these species are saturating all available niches, leading to greater evenness and higher prevalence (i.e., a larger fraction of the host population that is infected). These broad and opposing patterns call for future studies to investigate more directly what is driving geographic patterns of zoonotic host richness vs. richness of zoonotic diseases, an endeavor which subsumes direct confrontation of pervasive issues of sampling bias (for wildlife species) and reporting bias (for human disease cases) inherent to data on pathogens and disease.

Types of Pathogens

There are more zoonoses caused by bacteria than any other pathogen type, followed by viruses, helminths, protozoa, and fungi. While zoonotic pathogens are geographically widespread, they are unevenly distributed across mammal groups, with carnivores carrying the greatest number of viral and bacterial zoonoses, rodents carrying the most zoonotic helminths, and ungulates carrying more zoonotic protozoa compared to other mammal groups. Discerning the human disease risk posed by various pathogen types will require cross-referencing these broad patterns with zoonotic host patterns. For example, based purely on frequencies (i.e., excluding factors influencing contact rates with humans, etc.), the zoonotic virus hotspot in Europe could be driven by the high diversity of rodent hosts. Mapping zoonoses by pathogen type and mammal clade may better characterize the host and pathogen community contributing to broad patterns. Human disease risk posed by different pathogen types will also depend on their host breadth. In response to the emergence of prominent viral zoonoses in humans (e.g., MERS, SARS) and evidence of frequent host switching, recent studies have aimed to understand the zoonotic potential of viruses, particularly those arising from bats. In addition to characterizing the propensity for host switching among particular host clades and pathogen types, comparative studies of host competence will inform the relative contributions of host species as pathogen sinks or conduits for further transmission.

Primate Diseases

Due to the close genetic relationship between nonhuman primates and humans, disease causing organisms are easily exchanged between them. The pathogens that can be passed from nonhuman primates to humans and vice versa include bacteria, fungi, parasites, and viruses. They may be spread by bites, scratches, handling animals or their tissues, airborne transmission of aerosols and droplets, ingestion, and arthropod vectors. Often the nonhuman primate carries and transmits disease without any visible signs. Persons in contact with these animals must always be aware of the potential risks involved. This is especially true when animals are under stress, such as those that have been recently shipped or introduced into a new situation, or have developed a recent illness. As with many communicable diseases, immunocompromised persons are at greatest risk for infection or serious consequences from such infections.

Nonhuman primates can be divided into several groups; old world monkeys, new world monkeys and others. These groupings have significance when it comes to the kinds of diseases they can contract and transmit. For instance, one of the most serious diseases that humans can get from monkeys is caused by Herpesvirus simiae virus (B virus, Cercopithecine herpesvirus 1) that is enzootic among old world monkeys of the genus Macaca (macaques). Fortunately, most pet monkeys in this country are new world species and are bred in this country. Federal regulations prohibit importation of primates except for scientific, educational, and exhibition purposes. Importers must be registered with the Centers for Disease Control and Prevention and must hold a special permit to import cynomolgus, rhesus or African green monkeys. Unfortunately, animals may sometimes be smuggled into the country or adopted from research facilities.

Rabies

Nonhuman primate rabies is rarely reported in this country and with one exception has always occurred in animals that were recently imported from rabies endemic areas. The one exception occurred after a dog bit a pet monkey in 1911 during a dog rabies epidemic in Florida. Free ranging, macaque monkeys that have been introduced into Florida's Ocala Springs area, where raccoon rabies is endemic, have never been diagnosed with rabies. Between 1957 (the beginning of the raccoon rabies outbreak in Florida) and 1974 more than 640 nonhuman primates were tested for rabies in Florida.

B Virus (Cercopithecine Herpesvirus 1)

Up to 90% of adult macaques can be carriers of B-virus; most are asymptomatic, but some can have localized oral lesions. In humans the infection presents as a rapidly ascending encephalomyelitis with a fatality rate of about 70%. Most of the 25 well-documented cases of human infection have occurred in laboratory animal handlers who were somehow directly inoculated with tissue or fluid from a monkey via a bite, scratch, needlestick or laboratory injury. Although experimentally infected new world monkeys develop fatal disease and could conceivably become infected by contact with macaques, under most circumstances bites to humans from new world monkeys should not raise concern about this deadly disease because it is not endemic among new world monkeys.

Tuberculosis (TB)

Nonhuman primates are very susceptible to infection from mycobacteria tuberculosis (TB) and can contract it from humans or other animals. Primates from environments where human TB is prevalent are at greatest risk for having the disease. During the 31 day quarantine of legally imported primates, a minimum of three tuberculin skin tests are performed and positive animals are destroyed. Illegally imported monkeys and those raised and sold as pets in the US may not be appropriately tested and could be infected.

Local Wound Infections

Approximately 224 strains of bacteria have been identified in human and animal saliva-contaminated wounds. The organisms most often encountered in the mouths of rhesus monkeys are the Neisseria species, alpha hemolytic streptococci, and haemophilus parainfluenza. In addition, the attending physician should be notified of the possibility of infection with Eikenella corrodens, a facultative anaerobe associated with human and nonhuman primate bites, that cause extensive tissue damage.

Enteric Diseases

These are spread via the fecal oral route and cause similar symptoms in humans and nonhuman primates. The more common agents include bacteria (Shigella, Salmonella, Campylobacter), protozoan parasites (Cryptosporidium, Giardia, Amoeba, Balantidia), and helminth parasites (Strongyloides).

Simian Immunodeficiency Virus (SIV)

SIV is closely related to HIV-1 and HIV-2 (causes of AIDS) and causes an AIDS-like illness in macaque monkeys; it may be asymptomatic in other species. There have been no reports of human illness, but there are research workers who developed antibodies to SIV after handling laboratory specimens.

Marburg and Ebola (Filoviruses)

Humans have developed illness from Marburg infection when exposed to tissues from African Green monkeys. The Ebola viruses from the Sudan and Zaire have not been isolated from monkeys. A different Ebola virus was discovered in 1995 in West Africa chimpanzees when a researcher became infected. The Ebola virus that caused an outbreak in a Reston, Virginia monkey quarantine facility, did not cause illness in any humans, but four animal handlers developed antibodies to the virus. These incidents remind us of the potential for as yet undiscovered human pathogens to be introduced by wild caught monkeys.

Fibrosing Cardiomyopathy

Fibrosing cardiomyopathy is a disease commonly caused by a heart failure in great apes, most specially the males. When fibrosing cardiomyopathy attacks a healthy heart, it comes with a bacteria or a virus that makes the muscles of the heart turn into fibrous bands which makes them unable to pump blood in the blood streams. When a gorilla is stressed up, or the food it eats, then catecholamine which is a harmful substance is released in the heart muscle that make the C-reactive protein that is found in blood plasma produced by the liver to swell causing rheumatoid arthritis.

Contrast Analysis

Studies show that the causes of heart disease differ greatly between humans and chimpanzees. In this study, the scientists provided some new data and summarized existing reports on the subject. They also allow other primitives to have limited data, suggesting that they are more like chimpanzees in this respect. In general, the result is that heart disease does not represent a similarity between humans and other hominids, but rather an inexplicably special difference. Finally, the preliminary evidence of differences in extracellular matrix and glycosylation patterns between human and human-like hearts are proved and provided, which may be related to the understanding of these differences. Heart disease was the cause of 16 of the 52 deaths at Yakex primate research center between 1992 and 2005, and cardiac biopsies were carefully examined. This includes 9 animals (8 males, (1) dying females, (3) looking at the animals' serious animals (2 males, females). Almost all of these pathological abnormalities of death are associated with this type of FMI. Chimpanzees are very similar. An example shows a chimpanzee which goes through the heart muscle without hemiplegia, and heaven goes directly around the blood vessels, which can be seen in some people's hearts. For other reasons, the death of the Yerkes center also indicates that myocardial fibrosis was severe during this period of 14 men and 4 women, autopsies were fragmented by the IMF.

Food used to Alleviate Illness

Scientists have begun to study how billions of bacteria, fungi and other microbes living in the stomach and intestines of humans affect our health for the last few decades. What we eat determines which of these microorganisms thrive, and the composition of the intestinal flora has a great influence on other parts of our body. For example, some intestinal bacteria cause inflammation in our immune system, while other bacteria secrete substances that penetrate blood or block arteries, which helps explain why heart disease patients have different microbes and health conditions. Grains of paradise are plants that grow in swampy areas in West Africa-vine chocked swamps a member of the ginger family. It is a plant that gorillas like eating but it contains a powerful anti- swollen compound. It grows up to 1.5 meters with a trumpet shape and reddish-brown seeds. Gorillas do use the plant to make nests on the ground and beds that they use over the night for sleeping; they also use the seeds to treat coughs, toothaches and measles. The plant also provides comfort and warmth to the weak and cold bodies of the gorilla. The invention of processed high-calorie cookies containing vitamins and nutrients and the addition of several fruits and vegetables ultimately helped to standardize the diet of gorillas. The animal biscuit diet begins to prolong life and looks healthier and can sometimes survive for 50 years. The researches found that the biscuit diet has many shortcomings. Although gorillas are genetically similar to humans, their digestive systems are very different and more like horses. Like a horse, a gorilla is a digestive organ that processes food primarily in the very long large intestine, not in the stomach. This means they are good for breaking down the fiber, but not very good for sugar or grain. If the zookeeper feed them sweet potatoes or commercially grown fruit, they will eat them but that didn't bring much energy to them.

Category and Symptoms

There are different types of cardiomyopathy which include a hypertrophic cardiomyopathy which makes the heart muscles to enlarge and thicken; dilated cardiomyopathy happens when the ventricles enlarge and weaken; restrictive cardiomyopathy makes the ventricles to stiffen; Hypertrophic cardiomyopathy is an inherited one from one generation to another and dilated cardiomyopathy results due to heavy alcohol consumption, use of cocaine and viral infections. The signs and symptoms of cardiomyopathy include; shortness of breath, fatigue, swelling in the legs, dizziness, light-headedness, fainting during physical activities, irregular heartbeats, chest pain after heavy meals and unusual sounds associated with heartbeats. Gorillas inhabited the forests of central sub-Saharan Africa whereby they were divided into two species; the eastern gorillas and the western gorillas. They are much closer to humans because the DNA reveals a higher percentage between 95 to 99%. They fall under kingdom Mammalia same as the human and both have the same origin of common ancestors. Gorillas are considered to be a single species with three subspecies i.e. the western lowland gorilla, the eastern lowland gorilla and the mountain gorilla. Both the species became to be one after their forest habitat shrank and ended up separating. With gorillas that were captive by human, started developing fibrosing cardiomyopathy due to the foods that humans used to give them like biscuits diet which had lots of sugar and this made it difficult with digestion because of their hindgut digesters which processed food in their extra-long large intestines instead of their stomachs and had lesser energy distribution in their bodies. The new diet lowered the body fat and cholesterol and ended up affecting the bacteria living in gorillas' stomachs. A heart attack in humans occurs due to chest pain, sweating or even shortness of breath that results due to coronary artery having a problem in supplying blood into the heart muscle while with the gorillas, it happens due to the diet that the ones in captives used to take. Humans, who do not suffer from an acute coronary heart attack, do end up having a heart failure due to a gradual decrease of blood supply in the arteries. Both gorillas and humans have an unusual form of interstitial myocardial fibrosis whereby a normal myocardium in both humans and gorillas are quite similar to each other. The gorilla's heart fibrosis has been distributed in an unorderly manner in the cardiac muscle as seen in human.

Prevention

After so many attempts on how to prevent the fibrosing from attacking gorillas, the zookeepers came with the ideas of how they could reduce the mortality rate of the gorillas i.e. the introduction of a National Gorilla Cardiac Database which will be used in tracking cases of the disease to those gorillas that were in captivity in the western lowland; Introduction of a tab that determines the populations of the gorillas and also comparing the ultrasound waves that is to produce a visual display of the heart from a healthy gorilla to a sick gorilla so that they can detect the presence of the disease; Implantation of an advanced pacemaker in a gorilla that has the disease so that pacemaker can detect the disease at an early stage and also correcting the breakdown of the heart's electrical circuit that comes with the disease which later restores the heart to pump properly. Heart failure is considered to be common in both human and the gorillas which could be determined as a heart failure or a cardiac arrest to some point. When analysis is taken into an accord, a human heart attack would be considered to have occurred due to coronary artery atherosclerosis which happens when the arteries are hardened due to a buildup of plaque inside the walls of the arteries while for the gorillas it will be considered to have occurred due to the bacteria in the muscles of the heart that prevents the heart from pumping the blood properly into the arteries and the veins.

Horse Diseases

Horses are large, beautiful animals that people often own either as work animals or for casual riding. People who own horses often do so with a great sense of pride in having such a majestic creature. A part of this pride of ownership should come from providing horses with the care that they need to remain healthy. To successfully do that, it is crucial that horse owners and keepers recognize that there are numerous threats that can make a horse ill and even cause its death. Illness and disease can strike a horse without warning, and some can be fatal within a relatively short period of time. In addition to knowing of these illnesses, people must also be able to spot the signs that something is wrong, know what to do to provide immediate help to the animal, and know when to contact a vet for assistance. While there are many diseases that affect horses, some of the more common concerns include colic, equine arthritis, laminitis, West Nile Virus, equine encephalomyelitis, Potomac horse fever, azoturia, and botulism.

Colic

In horses, colic is a series of conditions that cause varying degrees of abdominal pain. It is a digestive disorder that is very common in horses. The severity of colic varies from mild to severe to the point of euthanization. The most common types of colic include spasmodic and impact colic. Spasmodic colic is caused by excessive gas that causes pain when it stretches the gut. Impact colic occurs when the gut is stretched due to a buildup of feed due to dryness or coarseness of the feed or some form of obstruction. Pain occurs when the stretched gut wall contracts in an attempt to push out the obstruction. Horses suffering from colic pain may bite at their flank or belly or make kicking motions toward it. In some cases, an animal may attempt to lie down or roll on the ground. Other symptoms of colic include anxiety, lack of appetite or defecation, playing in water, an elevated pulse rate, and seeming to play in their water bucket yet not drinking from it.

Degenerative Joint Disease (DJD)

Degenerative joint disease is a chronic and progressive type of arthritis. It is often called osteoarthritis and results in deterioration of the cartilage in the joints. The degree of joint damage associated with degenerative joint disease often leads to lameness. There is no cure for DJD, or osteoarthritis; however, affected animals may benefit from physical therapy or treatments for stiffness and pain in the joints using corticosteroids or NSAIDs.

Equine Arthritis

Equine arthritis is a term that is given to swelling, redness, and pain (inflammation) of the joints. The inflammation typically hinders the animal's ability to move comfortably and freely. There are several different types of arthritis that can affect horses, including osteoarthritis. These conditions include traumatic arthritis, septic arthritis, subchondral cystic lesions, and osteochondritis dissecans. Anti-inflammatory drugs can be helpful in the treatment of arthritis in aging horses, as may certain types of physical therapy.

Laminitis

Laminitis is an inflammation of the laminae in horse's hooves. The tissue is a type of connective tissue that attaches the coffin bone to the hoof wall. It is a very serious condition that can result in lameness in horses and may even lead to its eventual euthanasia. Typically, the condition affects the front hooves, but it can affect all four. The condition progresses through four stages, which include the developmental stage, acute, subacute, and chronic. Obesity, high fevers, and working on hard surfaces are considered to be risk factors.

Equine Encephalomyelitis

Mosquitoes carry the viruses that cause equine encephalomyelitis, which is also called sleeping sickness, a disease that affects the central nervous system. There are several strains of the virus, such as the St. Louis strain, the Western strain, and the Eastern strain. Horse owners can prevent it by vaccinating their horses and controlling mosquitoes in the area. Signs that a horse has this condition include depression, nervousness, fever, lack of coordination, poor reflexes, a drooping lower lip, and grinding of the teeth. Other severe symptoms involve the animal lying on its side while bicycling its legs, an inability to swallow, paralysis, and even death.

West Nile Virus

The West Nile Virus is a disease that horses get from being bitten by an infected mosquito. The virus can lead to inflammation of the brain, which is called encephalitis. Animals affected by West Nile Virus may also develop meningitis. Symptoms of a horse that is infected by West Nile include fever, impaired vision, convulsions, head pressing, difficulty swallowing, and paralysis or weakness of the hind limbs. Ideally, horse owners will want to vaccinate their horses to prevent infection and reduce areas that can attract mosquitoes, such as standing water.

Azoturia

Azoturia is a condition that affects a horse's muscles. It can cause problems such as cramping and stiffness. Often, this condition develops after a horse has been overexerted and is left to rest for a day without any changes to its diet. Azoturia has several names associated with it, including "Monday morning disease" and "tying up." In addition to the cramps and stiffness, a horse may stagger, have an elevated temperature, sweat, and have an elevated heart rate. The pain may be so bad that standing may be difficult or impossible for some horses. To treat this condition, seek the assistance of a vet and allow the animal to rest. Typically, the vet will give anti-inflammatories and muscle relaxants and require the muscles to be massaged. Recovery may take as long as eight weeks. People can prevent azoturia from occurring by adjusting the feed on days of rest and properly warming and cooling down their horses.

Botulism

There are three types of botulism that affects horses: botulism from spoiled hay that is either wet or dry, botulism from hay that is contaminated with the carcass of an animal, or botulism from a wound. Most often, horses that suffer from equine botulism get it from the hay that they eat. It is important for those caring for horses to recognize the signs of botulism, as an untreated horse is at high risk of dying or needing to be euthanized. The signs or symptoms associated with botulism include muscle weakness in the form of a weak tongue, weak eyelid tone, flaccid paralysis, and dysphagia. A horse that has come in contact with the toxin from contaminated feed may show edema of the face and muzzle, trembling of the muscles, and an inability to hold up its head. Treatment involves an antiserum.

Potomac Horse Fever

Potomac Horse Fever (PHF) is a potentially-fatal febrile illness affecting horses caused by the intracellular bacterium Neorickettsia risticii. PHF is also known as Shasta River Crud and Equine Monocytic Ehrlichiosis. It was first described in areas surrounding the Potomac River northwest of Washington, D.C., in the 1980s, but cases have been described in many other parts of the United States, such as Minnesota, California, and Pennsylvania. Currently, it is found in more than 40 U.S. states and Canada.

Cause

The causative agent of PHF is *Neorickettsia risticii* (formerly *Ehrlichia risticii*), an intracellular rickettsial bacterium.

Transmission

Accidental ingestion of the mayfly is thought to be one of the main modes of transmission of PHF.

The vector of *Neorickettsia risticii* is believed to be a trematode (fluke). The life cycle of the fluke takes it through freshwater snails and back into water, where it is ingested by the larval stages of several aquatic insects, including caddis flies and mayflies. It is thought that the main mode of infection is by accidental ingestion of infected adult insects, who may fly into barns and die in stalls or on pastures after enclosure. Experimental infection has been produced with oral administration of infected insects and subcutaneous inoculation of *N. risticii*. All attempts to transmit the disease using ticks have failed.

Several outbreaks of PHF have been found to coincide with mass emergences of burrowing mayflies of the genus *Hexagenia*; these insects hatch en masse and may be found littering the ground in nearby stables, where they are attracted by light. The entire natural history and life cycle of *N. risticii* has yet to be elucidated, but bats and birds may be wild reservoirs of infection. Unlike other causes of acute colitis in horses, such as *Salmonella* and *Clostridium*, PHF is not spread directly from horse to horse.

Signs and Symptoms

Signs and symptoms of PHF include acute-onset fever, depression (sometimes profound), inappetence, mild colic-like symptoms, decreased manure production, profuse watery non-fetid diarrhea endotoxemia, edema due to protein imbalances, abortion by pregnant mares, and acute laminitis (20 to 40 percent of cases). Infected horses founder usually within three days of the initial symptoms, thought to be secondary to endotoxemia. Death may occur and is usually due to severe laminitis leading to founder.

Horses may not always display any other symptoms beyond a fever.

Diagnosis

Diagnosis of PHF is accomplished by measuring antibody titers or PCR testing to look for the bacterium in the blood and feces. However, most veterinarians opt to initiate treatment right away, as the disease can progress quite quickly. Veterinarians may also run complete blood counts and chemistry and electrolyte panels to determine the course of care. Radiographs may be taken to track the progress of laminitic horses.

Treatment

N. risticii responds well to tetracycline antibiotics. Mild cases may be treated with oral doxycycline, while severe cases are usually treated with intravenous oxytetracycline.

Supportive care for severe cases is aimed at minimizing the effects of endotoxemia and preventing laminitis. This may include intravenous fluids and electrolytes to counteract the diarrhea; NSAIDs such as Banamine (flunixin meglumine); intravenous dimethyl sulfoxide; administration of products such as Biosponge or activated charcoal via nasogastric tube to bind endotoxins; polymyxin B or plasma for endotoxemia; supportive shoeing; low doses of intramuscular acepromazine; and pentoxifylline.

Prevention

While a vaccine is available for PHF, it does not cover all strains of the bacterium, and recent vaccine failures seem to be on the rise. Additionally, the vaccine usually produces a very weak immune response, which may only lessen the severity of the disease rather than prevent it. The vaccine is administered twice a year, in early spring and in early summer, with the first one inoculation given before the mayflies emerge and the second administered as a booster.

Some veterinarians have started making recommendations for farm management to try to prevent this disease:

- Maintaining riparian barriers along bodies of water may encourage aquatic insects to stay near their places of origin.

- Turning off outside lights around the barn will prevent insects from being attracted.

- Cleaning water buckets and feed areas frequently and keeping food covered will reduce the chance that the horse will accidentally ingest infected insects.

Skin Cancer in Horses

Skin cancer, or neoplasia, is the most common type of cancer diagnosed in horses, accounting for 45 to 80% of all cancers diagnosed. Sarcoids are the most common type of skin neoplasm and are the most common type of cancer overall in horses. Squamous-cell carcinoma is the second-most prevalent skin cancer, followed by melanoma. Squamous-cell carcinoma and melanoma usually occur in horses greater than 9-years-old, while sarcoids commonly affect horses 3 to 6 years old. Surgical biopsy is the method of choice for diagnosis of most equine skin cancers, but is contraindicated for cases of sarcoids. Prognosis and treatment effectiveness varies based on type of cancer, degree of local tissue destruction, evidence of spread to other organs (metastasis) and location of the tumor. Not all cancers metastasize and some can be cured or mitigated by surgical removal of the cancerous tissue or through use of chemotherapeutic drugs.

Sarcoids

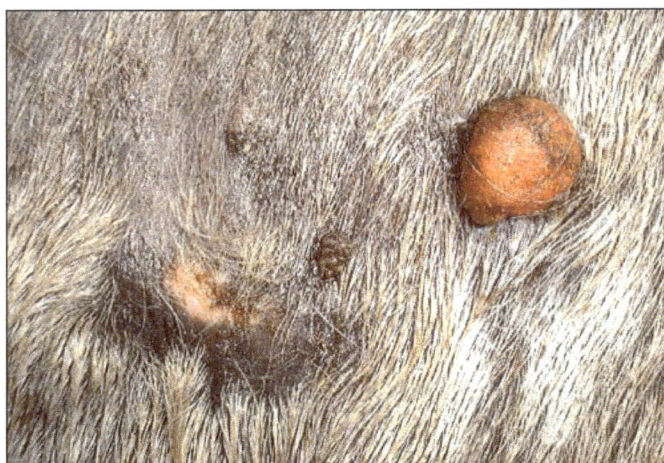

Occult (hairless area at left) and nodular (large round bump at right) forms of equine sarcoids.

Sarcoids account for 39.9% of all equine cancers and are the most common cancer diagnosed in horses. There is no breed predilection for developing sarcoids and they can occur at any age, with horses three to six years old being the most common age group and males being slightly more prone to developing the disease. Sarcoids are also more prevalent in certain familial lines, suggesting that there may be a heritable component. Several studies have found an association between the presence of Bovine papillomavirus-1 and 2 and associated viral growth proteins in skin cells with sarcoid formation, but the exact mechanism that controls or induces epidermal proliferation remains unknown. However, high viral loads within cells are strongly correlated with more severe clinical signs and aggressive lesions.

Clinical Signs

The appearance and number of sarcoids can vary, with some horses having single or multiple lesions, usually on the head, legs, ventrum and genitalia or around a wound. The distribution pattern suggests that flies are an important factor in the formation of sarcoids. Sarcoids may resemble warts (verrucous form), small nodules (nodular form), oval hairless or scaly plaques (occult form) or very rarely, large ulcerated masses (fibroblastic form). The occult form usually presents on skin around the mouth, eyes or neck, while nodular and verrucous sarcoids are common on the groin, penile sheath or face. Fibroblastic sarcoids have a predilection for the legs, groin, eyelid and sites of previous injury. Multiple forms may also be present on an individual horse (mixed form). Histologically, sarcoids are composed of fibroblasts (collagen producing cells) that invade and proliferate within the dermis and sometimes the subcutaneous tissue but do not readily metastasize to other organs. Surgical biopsy can definitively diagnose sarcoids, but there is a significant risk of making sarcoids worse. Therefore, diagnosis based solely on clinical signs, fine-needle aspiration or complete excisional biopsy are safer choices.

Treatment

While sarcoids may spontaneously regress regardless of treatment in some instances, course and duration of disease is highly unpredictable and should be considered on a case-by-case basis taking into account cost of the treatment and severity of clinical signs. Surgical removal alone is not effective, with recurrence occurring in 50 to 64% of cases, but removal is often done in conjunction with other treatments. Topical treatment with products containing bloodroot extract (from the plant *Sanguinaria canadensis*) for 7 to 10 days has been reported to be effective in removing small sarcoids, but the salve's caustic nature may cause pain and the sarcoid must be in an area where a bandage can be applied. Freezing sarcoids with liquid nitrogen (cryotherapy) is another affordable method, but may result in scarring or depigmentation. Topical application of the anti-metabolite 5-fluorouracil has also obtained favorable results, but it usually takes 30 to 90 days of repeated application before any effect can be realized. Injection of small sarcoids (usually around the eyes) with the chemotherapeutic agent cisplatin and the immunomodulator BCG have also achieved some success. In one trial, BCG was 69% effective in treating nodular and small fibroblastic sarcoids around the eye when repeatedly injected into the lesion and injection with cisplatin was 33% effective overall (mostly in horses with nodular sarcoids). However, BCG treatment carries a risk of allergic reaction in some horses and cisplatin has a tendency to leak out of sarcoids during repeated dosing. External beam radiation can also be used on small sarcoids, but is often impractical. Cisplatin electrochemotherapy (the application of an electrical field to the sarcoid after the injection of cisplatin, with the horse under general anesthesia), when used with or without prior surgery to remove the sarcoid, had a non-recurrence rate after four years of 97.9% in one retrospective study. There is a chance of sarcoid recurrence for all modalities even after apparently successful treatment. While sarcoids are not fatal, large aggressive tumors that destroy surrounding tissue can cause discomfort and loss of function and be resistant to treatment, making euthanasia justifiable in some instances. Sarcoids may be the most common skin-related reason for euthanasia.

Squamous-cell Carcinoma

Squamous-cell carcinoma (SCC) is the most common cancer of the eye, periorbital area and penis, and it is the second most common cancer overall in horses, accounting for 12 to 20% of all cancers diagnosed. While SCC has been reported in horses aged 1 to 29-years, most cases occur in 8 to 15-year-old horses, making it the most common neoplasm reported in older horses. Carcinomas are tumors derived from epithelial cells and SCC results from transformation and proliferation of squames, epidermal skin cells that become keratinized. Squamous-cell carcinomas are often solitary, slow-growing tumors that cause extensive local tissue destruction. They can metastasize to other organs, with reported rates as high as 18.6%, primarily to the lymph nodes and lung.

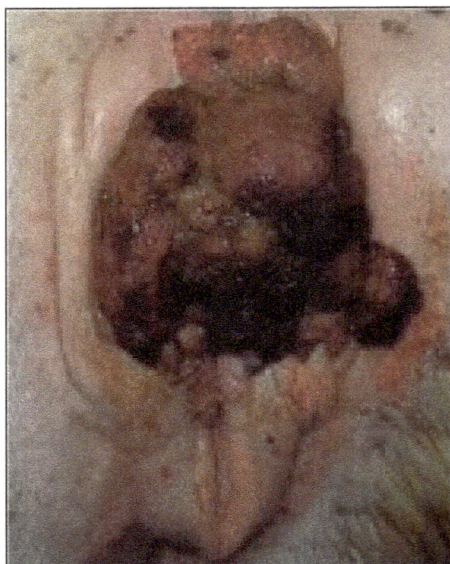

Squamous-cell carcinoma on the vulva of a gray mare, possibly arising from the clitoris. The tumor is ulcerated and has multiple necrotic areas (black spots).

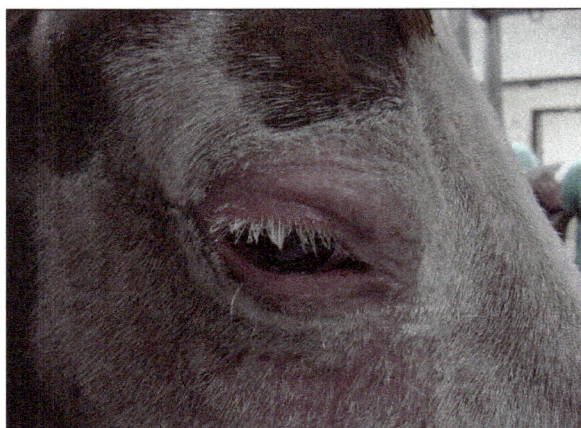

Smooth, raised plaque on upper eyelid of a Paint horse. This horse developed a carcinoma secondary to sunburn (termed solar keratosis carcinoma in situ).

Clinical Signs and Predisposing Factors

Tumors related to squamous-cell carcinoma (SCC) can appear anywhere on the body, but they are most often located in non-pigmented skin near mucocutaneous junctions (where skin meets mucous membranes) such as on the eyelids, around the nostrils, lips, vulva, prepuce, penis or anus. The tumors are raised, fleshy, often ulcerated or infected and may have an irregular surface. Rarely, primary SCC develops in the esophagus, stomach (non-glandular portion), nasal passages and sinuses, the hard palate, gums, guttural pouches and lung. The eyelid is the most common site, accounting for 40-50% of cases, followed by male (25-10% of cases) and female (10% of cases) genitalia. Horses with lightly pigmented skin, such as those with a gray hair coat or white faces, are especially prone to developing SCC, and some breeds, such as Clydesdales, may have a genetic predisposition. Exposure of light-colored skin to UV light has often been cited as a predisposing factor, but lesions can occur in dark skin and in areas that are not usually exposed to sunlight, such as around the anus. Buildup of smegma ("the bean" in horseman's terms) on the penis is also linked to SCC and is thought to be a carcinogen through penile irritation. Pony geldings and work horses are more prone to developing SCC on the penis, due to less frequent penile washing when compared to stallions. Equine papillomavirus-2 has also been found within penile SCCs, but has not been determined to cause SCC.

Treatment and Prevention of Scc

Before treatment of squamous-cell carcinoma (SCC) is initiated, evidence of metastasis must be determined either by palpation and aspiration of lymph nodes around the mass or, in smaller horses, radiographs of the thorax. Small tumors found early in the disease process (most frequently on the eyelid) can be treated with cisplatin or radiation with favorable results. For more advanced cases, surgical removal of eye (enucleation), mass or penile amputation can be curative provided all cancerous cells are removed (wide margins obtained) and there is no metastasis. However, young horses (usually geldings less than 8-years-old) that have a hard or "wooden" texture to SCCs on the glans penis have a very poor prognosis for treatment and recovery.

Regular washing of the penis and prepuce in males as well as cleaning the clitoral fossa (the groove around the clitoris) in mares is recommended to remove smegma buildup, which also gives the opportunity for inspection for suspicious growths on the penis or on the vulva.

Melanoma

Multiple nodules at the tail-base.

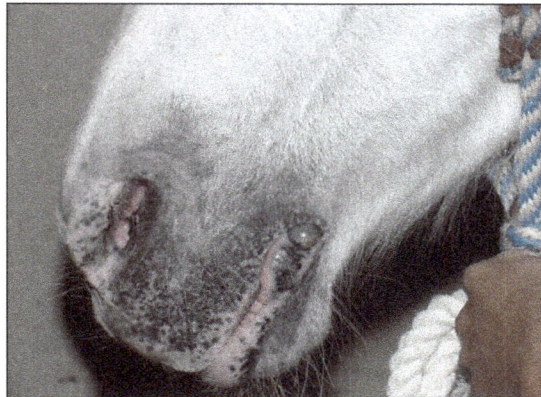

Small nodules at the lip commissure.

Common Sites for Melanoma

Equine melanoma results from abnormal proliferation and accumulation of melanocytes, pigmented cells within the dermis. Gray horses over 6-years-old are especially prone to developing melanoma. The prevalence of melanoma in gray horses over 15 years old has been estimated at 80%. One survey of Camargue-type horses found an overall population prevalence of 31.4%, with prevalence increasing to 67% in horses over 15 years old. Up to 66% of melanomas in gray horses are benign, but melanotic tumors in horses with darker hair-coats may be more aggressive and are more often malignant. One retrospective study of cases sent to a referral hospital reported a 14% prevalence of metastatic melanoma within the study population. However, the actual prevalence of metastatic melanoma may be lower due to infrequent submission of melanotic tumors for diagnosis. Common sites for metastasis include lymph nodes, the liver, spleen, lung, skeletal muscle, blood vessels and parotid salivary gland.

Clinical Signs

The most common sites for melanotic tumors are on the under-side of the tail near the base, on the prepuce, around the mouth or in the skin over the parotid gland (near the base of the ear). Tumors will initially begin as single, small raised areas that may multiply or coalesce into multi-lobed masses (a process called melanomatosis) over time. Horses under 2-years-old can be born with or acquire benign melanotic tumors (called melanocytomas), but these tumors are often located on the legs or trunk, not beneath the tail as in older animals.

Treatment of Melanoma

Treatment of small melanomas is often not necessary, but large tumors can cause discomfort and are usually surgically removed. Cisplatin and cryotherapy can be used to treat small tumors less than 3 centimeters, but tumors may reoccur. Cimetidine, a histamine stimulator, can cause tumors to regress in some horses, but may take up to 3 months to produce results and multiple treatments may be needed throughout the horse's life. There are few viable treatment options for horses with metastatic melanoma. However, gene therapy injections utilizing interleukin-12 and 18-encoding DNA plasmids have shown promise in slowing the progression of tumors in patients with metastatic melanoma.

Other Types of Skin Cancer

Lymphoma

Lymphoma is the most common type of blood-related cancer in horses and while it can affect horses of all ages, it typically occurs in horses aged 4–11 years.

African Horse Sickness

African horse sickness (AHS) is a highly infectious and deadly disease caused by African horse sickness virus. It commonly affects horses, mules, and donkeys. It is caused by a virus of the genus Orbivirus belonging to the family Reoviridae. This disease can be caused by any of the nine serotypes of this virus. AHS is not directly contagious, but is known to be spread by insect vectors.

AHS virus was first recorded south of the Sahara Desert in the mid-1600s, with the introduction of horses to southern Africa. The virus is considered endemic to the equatorial, eastern, and southern regions of Africa. Several outbreaks have occurred in the Equidae throughout Africa and elsewhere. AHS is known to be endemic in sub-Saharan Africa, and has spread to Morocco, the Middle East, India, and Pakistan. More recently, outbreaks have been reported in the Iberian Peninsula. AHS has never been reported in the Americas, eastern Asia, or Australasia. Epidemiology is dependent on host-vector interaction, where cyclic disease outbreaks coincide with high numbers of competent vectors. The most important vector for AHS in endemic areas is the biting midge

Culicoides imicola, which prefers warm, humid conditions. Larvae do not carry the virus, and long, cold winters are sufficient to break epidemics in nonendemic areas.

Host

The common hosts of this disease are horses, mules, donkeys, and zebras. However, elephants, camels, and dogs can be infected, as well, but often show no signs of the disease. Dogs usually contract the disease by eating infected horse meat, a recent report has been made of the disease occurring in dogs with no known horse-meat ingestion.

Transmission

This disease is spread by insect vectors. The biological vector of the virus is the *Culicoides* (midges) species. However, this disease can also be transmitted by species of mosquitoes including *Culex*, *Anopheles*, and *Aedes*, and species of ticks such as *Hyalomma* and *Rhipicephalus*.

Clinical Signs

Horses are the most susceptible host with close to 90% mortality of those affected, followed by mules (50%) and donkeys (10%). African donkeys and zebras very rarely display clinical symptoms, despite high virus titres in blood, and are thought to be the natural reservoir of the virus. AHS manifests itself in four different forms.

Pulmonary Form

The peracute form of the disease is characterized by high fever, depression, and respiratory symptoms. The clinically affected animal has trouble breathing, starts coughing frothy fluid from nostril and mouth, and shows signs of pulmonary edema within four days. Serious lung congestion causes respiratory failure and results in death in under 24 hours. This form of the disease has the highest mortality rate.

Cardiac Form

This subacute form of the disease has an incubation period longer than that of the pulmonary form. Signs of disease start at day 7–12 after infection. High fever is a common symptom. The disease also manifests as conjunctivitis, with abdominal pain and progressive dyspnea. Additionally, edema is presented under the skin of the head and neck, most notably in swelling of the supraorbital fossae, palpebral conjunctiva, and intermandibular space. Mortality rate is between 50 and 70%, and survivors recover in 7 days.

Mild or Horse Sickness Fever Form

Mild to subclinical disease is seen in zebras and African donkeys. Infected animals may have a low-grade fever and congested mucous membrane. The survival rate is 100%.

Mixed Form

Diagnosis is made at necropsy. Affected horses show signs of both the pulmonary and cardiac forms of AHS.

Diagnosis

Presumptive diagnosis is made by characteristic clinical signs, *post* mortem lesions, and presence of competent vectors. Laboratory confirmation is by viral isolation, with such techniques as quantitative PCR for detecting viral RNA, antigen capture (ELISA), and immunofluorescence of infected tissues. Serological tests are only useful for detecting recovered animals, as sick animals die before they are able to mount effective immune responses.

Treatment and Prevention

No treatment for AHS is known.

Control of an outbreak in an endemic region involves quarantine, vector control, and vaccination. To prevent this disease, the affected horses are usually slaughtered, and the uninfected horses are vaccinated against the virus. Three vaccines currently exist, which include a polyvalent vaccine, a monovalent vaccine, and a monovalent inactivated vaccine. This disease can also be prevented by destroying the insect vector habitats and by using insecticides.

African horse sickness was diagnosed in Spain in 1987–90 and in Portugal in 1989, but was eradicated using slaughter policies, movement restrictions, vector eradication, and vaccination.

AHS is related to bluetongue disease and is spread by the same midge (*Culicoides species*).

Dog Diseases

Infectious Canine Hepatitis

Infectious canine hepatitis (ICH) is an acute liver infection in dogs caused by Canine mastadenovirus A, formerly called Canine adenovirus 1 (CAV-1). CAV-1 also causes disease in wolves, coyotes, and bears, and encephalitis in foxes. The virus is spread in the feces, urine, blood, saliva, and nasal discharge of infected dogs. It is contracted through the mouth or nose, where it replicates in the tonsils. The virus then infects the liver and kidneys. The incubation period is 4 to 7 days.

Symptoms include fever, depression, loss of appetite, coughing, and a tender abdomen. Corneal edema and signs of liver disease, such as jaundice, vomiting, and hepatic encephalopathy, may also occur. Severe cases will develop bleeding disorders, which can cause hematomas to form in the mouth. Death can occur secondary to this or the liver disease. However, most dogs recover after a brief illness, although chronic corneal edema and kidney lesions may persist.

Diagnosis is made by recognizing the combination of symptoms and abnormal blood tests that occur in infectious canine hepatitis. A rising antibody titer to CAV-1 is also seen. The disease can be confused with canine parvovirus because both will cause a low white blood cell count and bloody diarrhea in young, unvaccinated dogs.

Treatment is symptomatic. Most dogs recover spontaneously without treatment. Prevention is through vaccination (ATCvet code QI07AA05 (WHO) and various combination vaccines).

Most combination vaccines for dogs contain a modified canine adenovirus type-2. CAV-2 is one of the causes of respiratory infections in dogs, but it is similar enough to CAV-1 that vaccine for one creates immunity for both. CAV-2 vaccine is much less likely to cause side effects than CAV-1 vaccine. One study has shown the vaccine to have a duration of immunity of at least four years.

CAV-1 is destroyed in the environment by steam cleaning and quaternary ammonium compounds. Otherwise, the virus can survive in the environment for months in the right conditions. It can also be released in the urine of a recovered dog for up to a year.

White Dog Shaker Syndrome

White dog shaker syndrome (also known as idiopathic steroid responsive shaker syndrome, shaker dog syndrome and "little white shakers" syndrome; Latin name Idiopathic Cerebellitis) causes full body tremors in small dog breeds. It is most common in West Highland White Terriers, Maltese, Bichons, and Poodles, and other small dogs. There is a sudden onset of the disease at one to two years of age. It is more likely to occur, and the symptom is worse during times of stress. Nystagmus, difficulty walking, and seizures may occur in some dogs.

The cause is unknown, but it may be mediated by the immune system. One theory is that there is an autoimmune-induced generalized deficiency of neurotransmitters. Cerebrospinal fluid analysis may reveal an increased number of lymphocytes. Treatment with corticosteroids may put the dog into remission, or diazepam may control the symptoms. Typically the two drugs are used together. There is a good prognosis, and symptoms usually resolve with treatment within a week, although lifelong treatment may be necessary.

Kennel Cough

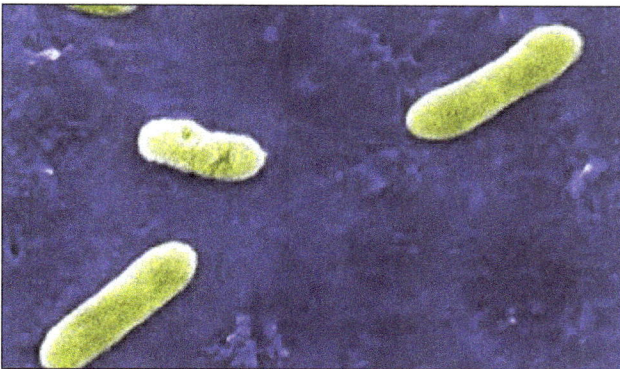

A scanning electron micrograph (SEM) depicting a number of Gram-negative *Bordetella bronchiseptica* bacteria.

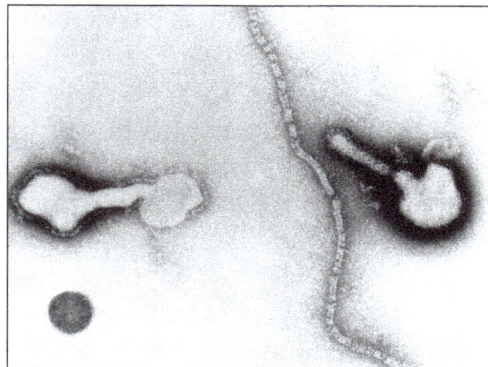

Transmission electron micrograph of parainfluenza virus. Two intact particles and free filamentous nucleocapsid.

Kennel cough also known as canine infectious respiratory disease, formerly canine infectious tracheobronchitis is an upper respiratory infection affecting dogs. There are multiple causative agents, the most common being the bacterium Bordetella bronchiseptica (found in 78.7% of cases in Southern Germany), followed by canine parainfluenza virus (37.7% of cases), and to a lesser extent canine coronavirus (9.8% of cases). It is highly contagious; however adult dogs may display immunity to reinfection even under constant exposure. Kennel cough is so named because the infection can spread quickly among dogs in the close quarters of a kennel or animal shelter.

Viral and bacterial causes of canine cough are spread through airborne droplets produced by sneezing and coughing. These agents also spread through contact with contaminated surfaces. Symptoms begin after a several day incubation period post-exposure, and in most cases will clear up on their own. However, in young puppies or immunocompromised animals, mixed or secondary infections can progress to lower respiratory infections such as pneumonia.

Symptoms

The incubation period is 5–7 days (with a range of 3–10). Symptoms can include a harsh, dry cough, retching, sneezing, snorting, gagging or vomiting in response to light pressing of the trachea or after excitement or exercise. The presence of a fever varies from case to case.

Types

Although kennel cough is considered to be a multifactorial infection, there are two main forms. The first is more mild and is caused by *B. bronchiseptica* and canine parainfluenza virus infections, without complications from canine distemper virus (CDV) or canine adenovirus (CAV). This form occurs most regularly in autumn, and can be distinguished by symptoms such as a retching cough and vomiting. The second form has a more complex combination of causative organisms including CDV and CAV. It typically occurs in dogs that have not been vaccinated and it is not seasonal. Symptoms are more severe than the first form, and may include rhinitis, conjunctivitis, and fever in addition to a hacking cough.

Transmission

Viral infections such as canine parainfluenza or canine coronavirus are only spread for roughly one week following recovery; however, respiratory infections involving *B. bronchiseptica* can be transmissible for several weeks longer. While there was early evidence to suggest that *B. bronchiseptica* could be shed for many months post-infection, a more recent report places detectable nasal and pharyngeal levels of *B. bronchiseptica* in 45.6% of all clinically healthy dogs. This has potentially expanded the vector from currently or recently infected dogs to half the dog population as carriers. To put the relative levels of shedding bacteria into perspective, a study analyzing the shedding kinetics of *B. bronchiseptica* presents the highest levels of bacterial shedding one week post-exposure, with an order of magnitude decrease in shedding observed every week. This projection places negligible levels of shedding to be expected six weeks post-exposure (or approximately five weeks post-onset of symptoms). Dogs which had been administered intranasal vaccine four weeks prior to virulent *B. bronchiseptica* challenge displayed little to no bacterial shedding within three weeks of exposure to the virulent strain.

Treatment and Prevention

Antibiotics are given to treat any bacterial infection present. Cough suppressants are used if the cough is not productive. NSAIDs are often given to reduce fever and upper respiratory inflammation. Prevention is by vaccinating for canine adenovirus, distemper, parainfluenza, and *Bordetella*.

In kennels, the best prevention is to keep all the cages disinfected. In some cases, such as "doggie daycares" or nontraditional playcare-type boarding environments, it is usually not a cleaning or disinfecting issue, but rather an airborne issue, as the dogs are in contact with each other's saliva and breath. Although most kennels require proof of vaccination, the vaccination is not a fail-safe preventative. Just like human influenza, even after receiving the vaccination, a dog can still contract mutated strains or less severe cases.

Vaccines

To increase their effectiveness, vaccines should be administered as soon as possible after a dog enters a high-risk area, such as a shelter. 10 to 14 days are required for partial immunity to develop. Administration of *B. bronchiseptica* and canine-parainfluenza vaccines may then be continued routinely, especially during outbreaks of kennel cough. There are several methods of administration, including parenteral and intranasal. However, the intranasal method has been recommended when exposure is imminent, due to a more rapid and localized protection. Several intranasal vaccines have been developed that contain canine adenovirus in addition to *B. bronchiseptica* and canine parainfluenza virus antigens. Studies have thus far not been able to determine which formula of vaccination is the most efficient. Adverse effects of vaccinations are mild, but the most common effect observed up to 30 days after administration is nasal discharge. Vaccinations are not always effective. In one study it was found that 43.3% of all dogs in the study population with respiratory disease had in fact been vaccinated.

Complications

Dogs will typically recover from kennel cough within a few weeks. However, secondary infections could lead to complications that could do more harm than the disease itself. Several opportunistic invaders have been recovered from the respiratory tracts of dogs with kennel cough, including *Streptococcus*, *Pasteurella*, *Pseudomonas*, and various coliform bacteria. These bacteria have the potential to cause pneumonia or sepsis, which drastically increase the severity of the disease. These complications are evident in thoracic radiographic examinations. Findings will be mild in animals affected only by kennel cough, while those with complications may have evidence of segmental atelectasis and other severe side effects.

Hypertrophic Osteodystrophy

Hypertrophic Osteodystrophy (HOD) is a bone disease that occurs in fast-growing large and giant breed dogs. The disorder is sometimes referred to as metaphyseal osteopathy, and typically first presents between the ages of 2 and 7 months. HOD is characterized by decreased blood flow to the metaphysis (the part of the bone adjacent to the joint) leading to a failure of ossification (bone formation) and necrosis and inflammation of cancellous bone. The disease is usually bilateral in the limb bones, especially the distal radius, ulna, and tibia.

The Weimaraner, Irish Setter, Boxer, German Shepherd, and Great Dane breeds are heavily represented in case reports of HOD in the veterinary literature, but the severity of symptoms and possible etiology may be different across the breeds. For example, familial clustering of the disease has been documented in the Weimaraner, but not in other breeds. The disease in the

Weimaraner and Irish Setter can be particularly severe, with significant mortality observed in untreated dogs. The classical age of onset is typically 8 to 16 weeks of age, with males and females equally affected.

Speculated Causes of HOD

Causes have been speculated to include decreased Vitamin C uptake, increased vitamin (other than C) and mineral uptake, and infection with canine distemper virus(CDV). Decreased Vitamin C uptake has been dismissed as a cause, but excessive calcium supplementation remains a possibility. There is no evidence over-feeding is a significant cause. In Weimaraners, recent vaccination with a modified live vaccine has been a suspected cause, partly because HOD often presents immediately after a vaccination, and partly because of the autoimmune nature of the disorder. The canine distemper vaccination in particular has been a suspected causal factor due to the significant number of overlapping symptoms observed between systemically affected HOD puppies and dogs suffering from distemper, but to-date, no definitive linkage has been demonstrated. The cause of canine HOD largely remains unknown. However, because of the familial clustering, HOD in the Weimaraner is suspected to have a genetic, or partly genetic, origin.

Clinical Features

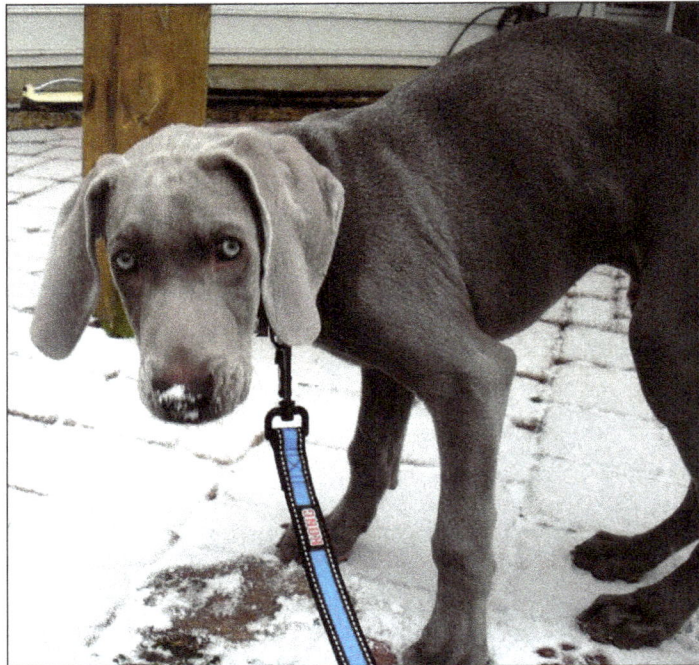

Characteristic roaching in 4-month-old Weimaraner puppy with HOD.

A primary characteristics of the condition is lameness. This is generally due to the swelling of the metaphysis of the long bones that is observed. Other bones may be affected, particularly the ribs, the metacarpal bones, the mandible, and the scapula. Lameness may present as mild limping or more severely as a reluctance or inability to stand. In some breeds and individuals, the stance of an HOD puppy as observed from behind has sometimes been described as "cowhocked." Shaking of limbs and a reluctance to put full body weight on the front legs is often observed. Sometimes the puppy will exhibit a characteristic "roaching" or arching of the spine when standing.

Lameness is accompanied by pain upon palpation of affected bones, warmth in the limb as felt by the inside of the clinician's wrist, depression, and loss of appetite. Limb involvement is usually bilateral, typically involves the distal radius and ulna, and may be episodic. There is evidence to suggest that most dogs recover after one episode, but some relapse. Dogs suffering systemic manifestations of the disorder often have poorer prognoses. Systemic manifestations include fever, multiple body organ inflammation, nasal and ocular discharge, Diarrhea, hyperkeratosis of the foot pads, pneumonia, and tooth enamel hypoplasia (many of these symptoms overlap with symptoms of CDV). Because early diagnosis must be based on clinical signs, to the extent possible, other disorders should be ruled out (e.g., nutritional bone disease, and osteochondrosis).

Radiographic Features

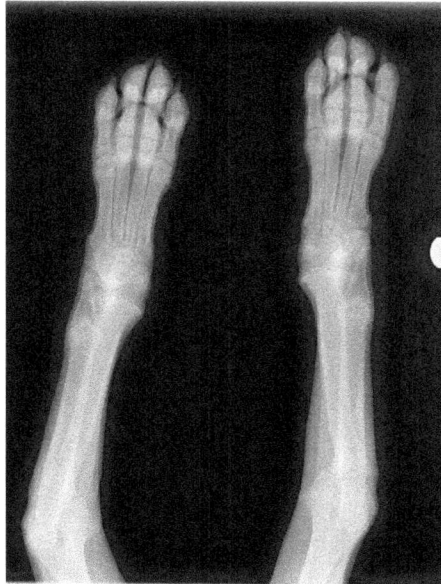

X-Ray image of HOD presentation in 4-month-old Weimaraner puppy.
"Moth-eaten" appearance of metaphyses.

Diagnosis relies on clinical signs and characteristic changes in radiographic images of the metaphyses. Bone changes can be observed on radiograph, and the disorder may progress to actual angular limb deformity. In the early stage of the illness, the metaphyseal area on X-ray may be observed to have an uneven radiolucent zone parallel to the physis with a thin band of increased radiodensity directly bordering the physis. Early stage radiographic changes have sometimes been described as having a "moth-eaten" appearance. As the disease progresses, the radiolucent line may disappear and radiodensity may increase in the affected area as the body attempts to repair damage. Relapses can cause new radiolucent lines. This area is often followed by a dark line at the metaphysis, which may progress to new bone growth on the outside of that area. This area represents microfractures in the metaphysis and bone proliferation to bridge the defect in the periosteum.

Treatment

Treatment options have been controversial. Mild illness is often successfully treated with pain medication (usually NSAIDs) and supportive care. Dogs presenting with severe, systemic symptoms not responding to NSAID treatment require more intensive treatment. The Weimaraner and Irish Setter American Kennel Club (AKC) parent clubs advocate the use of immunosuppressive doses of corticosteroids, supplemented with antibiotics and antacids (to compensate for the decreased thickness of the stomach's mucosal lining as a result of the corticosteroids and to decrease the possibility of forming stomach ulcers). More severe cases that are not recognized and treated early often require IV fluids, electrolytes, nutritional support, and significant nursing care to achieve successful results. AKC parent clubs have supported research into the genetic causes of HOD, and have reported fairly good success using this protocol, saving many puppies from unnecessary suffering, deformity, and death.

Dog Skin Disorders

Skin disorders are among the most common health problems in dogs, and have many causes. The condition of a dog's skin and coat are also an important indicator of its general health. Skin disorders of dogs vary from acute, self-limiting problems to chronic or long-lasting problems requiring life-time treatment. Skin disorders may be primary or secondary (due to scratching, itch) in nature, making diagnosis complicated.

Immune-mediated Skin Disorders

Skin disease may result from deficiency or overactivity of immune responses. In cases where there are insufficient immune responses, the disease is usually described by the secondary disease that results. Examples include increased susceptibility to demodectic mange and recurrent skin infections, such as Malassezia infection or bacterial infections. Increased but harmful immune responses can be divided into hypersensitivity disorders such as atopic dermatitis and autoimmune disorders (autoimmunity), such as pemphigus and discoid lupus erythematosus.

Atopic Dermatitis

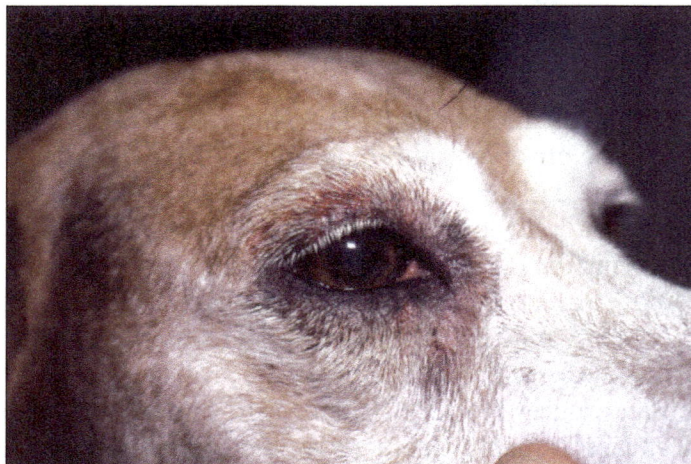

Dog with atopic dermatitis, with signs around the eye created by rubbing.

Atopy is a hereditary and chronic (lifelong) allergic skin disease. Signs usually begin between 6 months and 3 years of age, with some breeds of dog, such as the Golden Retriever, showing signs at an earlier age. Dogs with atopic dermatitis are itchy, especially around the eyes, muzzle, ears and feet. In severe cases, the irritation is generalised. If the allergens are seasonal, the signs of irritation are similarly seasonal. Many dogs with house dust mite allergy have perennial disease. Some of the allergens associated with atopy in dogs include pollens of trees, grasses and weeds, as well as molds and House dust mites. Ear and skin infections by the bacteria *Staphylococcus pseudintermedius* and the yeast *Malassezia pachydermatis* are common secondary to atopic dermatitis.

Food allergy can be associated with identical signs and some authorities consider food allergy to be a type of atopic dermatitis. Food allergy can be identified through the use of elimination diet trials in which a novel or hydrolysed protein diet is used for a minimum of 6 weeks.

Diagnosis of atopic dermatitis is by elimination of other causes of irritation, including fleas, mites, and other parasites, such as *Cheyletiella* and lice. Allergies to aeroallergens can be identified using intradermal allergy testing and blood testing (allergen-specific IgE ELISA).

Treatment includes avoidance of the offending allergens if possible, but for most dogs this is not practical or effective. Other treatments modulate the adverse immune response to allergens and include antihistamines, steroids, ciclosporin, and immunotherapy (a process in which allergens are injected to try to induce tolerance). In many cases, shampoos, medicated wipes and ear cleaners are needed to try to prevent the return of infections.

Autoimmune Skin Diseases

Pemphigus foliaceus is the most common autoimmune disease of the dog. Blisters in the epidermis rapidly break to form crusts and erosions, most often affecting the face and ears initially, but in some cases spreading to include the whole body. The paw pads can be affected, causing marked hyperkeratosis (thickening of the pads with scale). Other autoimmune diseases include bullous pemphigoid and epidermolysis bullosa acquisita.

Treatment of autoimmune skin requires methods to reduce the abnormal immune response; steroids, azathioprine and other drugs are used as immunosuppressive agents.

Physical and Environmental Skin Diseases

Hot Spots

A hot spot, or acute moist dermatitis, is an acutely inflamed and infected area of skin irritation created and made worse by a dog licking and biting at itself. A hot spot can manifest and spread rapidly in a matter of hours, as secondary *Staphylococcus* infection causes the top layers of the skin to break down and pus becomes trapped in the hair. Hot spots can be treated with corticosteroid medications and oral or topical antibiotic applications, as well as clipping hair from around the lesion. Underlying causes include flea allergy dermatitis, ear infections, or other allergic skin diseases. Dogs with thick undercoats are most susceptible to developing hot spots.

Acral Lick Granulomas

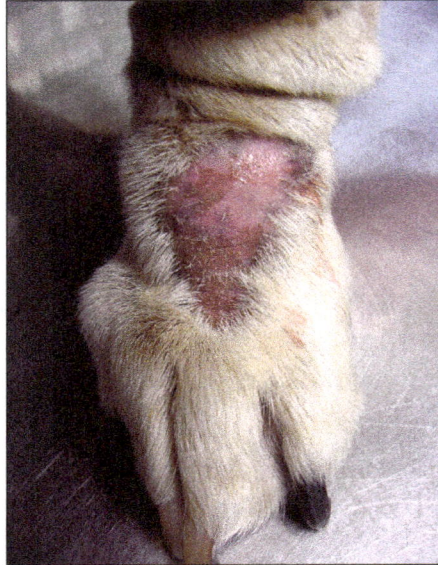

Lick granuloma from excessive licking.

Lick granulomas are raised, usually ulcerated areas on a dog's extremity caused by the dog's own incessant, compulsive licking. Compulsive licking is defined as licking in excess of that required for standard grooming or exploration, and represents a change in the animal's typical behavior and interferes with other activities or functions (e.g., eating, drinking, playing, interacting with people) and cannot easily be interrupted.

Infectious Skin Diseases

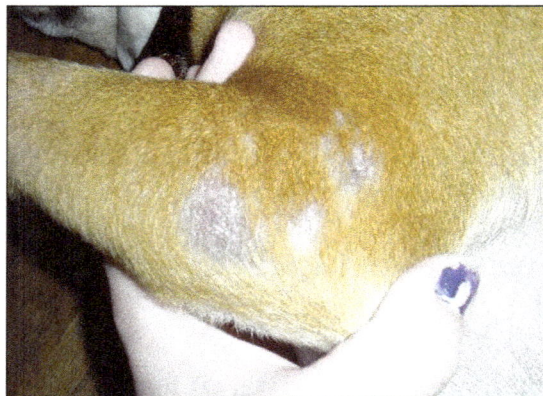

A dog with skin irritation and hair loss on its leg caused by demodectic mange.

Infectious skin diseases of dogs include contagious and non-contagious infections or infestations. Contagious infections include parasitic, bacterial, fungal and viral skin diseases.

One of the most common contagious parasitic skin diseases is Sarcoptic mange (scabies). Another is mange caused by Demodex mites (Demodicosis), though this form of mange is not contagious. Another contagious infestation is caused by a mite, *Cheyletiella*. Dogs can be infested with contagious lice.

Other ectoparasites, including flea and tick infestations are not considered directly contagious but are acquired from an environment where other infested hosts have established the parasite's life cycle.

Ringworm is a fungal skin infection and is more common in puppies than in adult dogs.

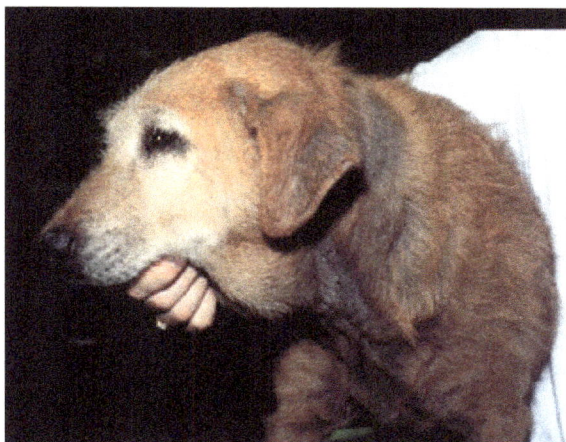
Dog with dermatitis caused by Malassezia (yeast).

Non-contagious skin infections can result when normal bacterial or fungal skin flora is allowed to proliferate and cause skin disease. Common examples in dogs include *Staphylococcus intermedius pyoderma*, and *Malassezia dermatitis* caused by overgrowth of *Malassezia pachydermatis*.

Alabama rot, which is believed to be caused by *E. coli* toxins, also causes skin lesions and eventual kidney failure in 25% of cases.

Hereditary and Developmental Skin Diseases

Some diseases are inherent abnormalities of skin structure or function. These include seborrheic dermatitis, ichthyosis, skin fragility syndrome (Ehlers-Danlos), hereditary canine follicular dysplasia and hypotrichosis, such as color dilution alopecia.

Juvenile cellulitis, also known as puppy strangles, is a skin disease of puppies of unknown etiology, which most likely has a hereditary component related to the immune system.

Cutaneous Manifestations of Internal Diseases

Some systemic diseases can become symptomatic as a skin disorder. These include many endocrine (hormonal) abnormalities, such as hypothyroidism, Cushing's syndrome (hyperadrenocorticism), and tumors of the ovaries or testicles.

Nutritional Basis of Skin Disorders

Essential Fatty Acids

Many canine skin disorders can have a basis in poor nutrition. The supplementation of both omega fatty acids 3 and 6 have been shown to mediate the inflammatory skin response seen in chronic diseases. Omega 3 fatty acids are increasingly being used to treat pruritic, irritated skin. A group of dogs supplemented with omega 3 fatty acids (660 mg/kg [300 mg/lb] of body weight/d) not only improved the condition of their pruritus, but showed an overall improvement in skin condition. Furthermore, diets lacking in essential fatty acids usually present as matted and unkept fur as the first sign of a deficiency. Eicosapentaenoic acid (EPA), a well known omega 3, works by preventing the synthesis of another omega metabolite known as arachidonic acid. Arachidonic acid is an omega 6, making it pro-inflammatory. Though not always the case, omega 6 fatty acids promote inflammation of the skin, which in turn reduces overall appearance and health. There are skin benefits of both these lipids, as a deficiency in omega 6 leads to a reduced ability to heal and a higher risk of infection, which also diminishes skin health. Lipids in general benefit skin health of dogs, as they nourish the epidermis and retain moisture to prevent dry, flaky skin.

Vitamins

Vitamins are one of many of the nutritional factors that change the outward appearance of a dog. The fat soluble vitamins A and E play a critical role in maintaining skin health. Vitamin A, which can also be supplemented as beta-carotene, prevents the deterioration of epithelial tissues associated with chronic skin diseases and aging. A deficiency in vitamin A can lead to scaly of skin and other dermatitis-related issues like alopecia. Vitamin E is an antioxidant. Vitamin E neutralizes free radicals that accumulate in highly proliferative cells like skin and prevent the deterioration of fibrous tissue caused by these ionized molecules. There are also a couple of water-soluble vitamins that contribute to skin health. Riboflavin (B2) is a cofactor to the metabolism of carbohydrates and when deficient in the diet leads to cracked, brittle skin. Biotin (B7) is another B vitamin that, when deficient, leads to alopecia.

Minerals

Minerals have many roles in the body, which include acting as beneficial antioxidants. Selenium is an essential nutrient, that should be present in trace amounts in the diet. Like other antioxidants, selenium acts as a cofactor to neutralize free radicals. Other minerals act as essential cofactors to biological processes relating to skin health. Zinc plays a crucial role in protein synthesis, which aids in maintaining elasticity of skin. By including zinc in the diet it will not only aid in the development of collagen and wound healing, but it will also prevent the skin from becoming dry and flaky. Copper is involved in multiple enzymatic pathways. In dogs, a deficiency in copper results in incomplete keratinization leading to dry skin and hypopigmentation. The complicated combination of trace minerals in the diet are a key component of skin health and a part of a complete and balanced diet.

Cat Diseases

Congenital Sensorineural Deafness in Cats

Congenital sensorineural deafness occurs in domestic cats with a white coat. It is a congenital deafness caused by a degeneration of the inner ear. Deafness can occur in white cats with yellow, green or blue irises, although it is mostly likely in white cats with blue irises. In white cats with mixed-coloured eyes (odd-eyed cats), it has been found that deafness is more likely to affect the ear on the blue-eyed side. White cats can have blue, gold, green, or copper coloured odd eyes.

In one 1997 study of white cats with varying degrees of hearing deficiency, 72% of the animals were found to be totally deaf. The entire organ of Corti was found to have degenerated within the first few weeks after birth; however, even during these weeks no brain stem responses could be evoked by auditory stimuli, suggesting that these animals had never experienced any auditory sensations. It was found that some months after the organ of Corti had degenerated, the spiral ganglion also began to degenerate.

Domesticated white cats with blue eyes and white coats are often completely deaf. Whether or not this is a result of Waardenburg syndrome remains unclear. Deafness is far more common in white cats than in those with other coat colors. According to the ASPCA Complete Guide to Cats, "17 to 22 percent of white cats with nonblue eyes are deaf; 40 percent of "odd-eyed" white cats with one blue eye are deaf; and 65 to 85 percent of blue-eyed white cats are deaf."

Genetics

The gene that causes a cat to have a white coat is a dominant masking gene. As a result, the cat will have an underlying coat colour and pattern. When the dominant white gene is present, however, that pattern will not be expressed. A cat that is homozygous (WW) or heterozygous (Ww) for this gene will have a white coat despite the underlying pattern/colour. A cat that lacks this dominant masking gene (ww) will exhibit a coat colour/pattern. There are several sources for a white cat to have blue eyes. If the underlying coat pattern is one of a pointed cat (also referred to as a Siamese pattern), the blue eyes may come from the genetics of the pointed gene. A common misconception is that all white cats with blue eyes are deaf. It is possible to have a cat with a naturally white coat without this gene, as an extreme form of white spotting, although this is rare some small non-white patch usually remains.

A completely deaf, pure white, blue-eyed cat.

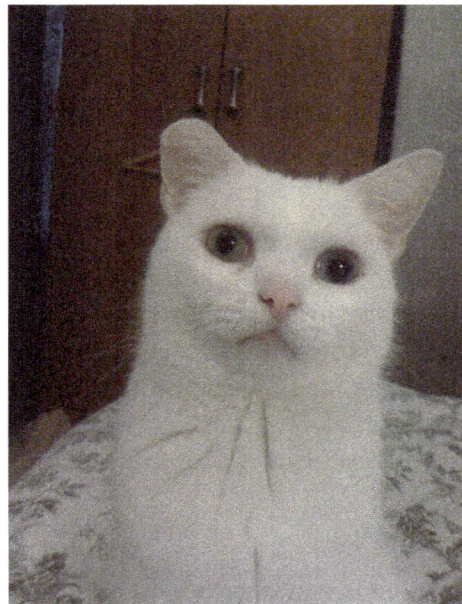

Deaf odd-eyed white cat.

Diabetes in Cats

Diabetes mellitus is a chronic disease in cats, whereby either insufficient insulin response or insulin resistance lead to persistently high blood glucose concentrations. Diabetes could affect up to 1 in 230 cats, and may be becoming increasingly common. Diabetes mellitus is less common in cats than in dogs. 80-95% of diabetic cats experience something similar to type-2 diabetes, but are generally severely insulin-dependent by the time symptoms are diagnosed. The condition is treatable, and if treated properly, the cat can experience a normal life expectancy. In type-2 cats, prompt effective treatment may lead to diabetic remission, in which the cat no longer needs injected insulin. Untreated, the condition leads to increasingly weak legs in cats, and eventually malnutrition, ketoacidosis and dehydration, and death.

Symptoms

Cats will generally show a gradual onset of the disease over a few weeks or months, and it may escape notice for even longer.

The first outward symptoms are a sudden weight loss (or occasionally gain), accompanied by excessive drinking and urination; for example, cats can appear to develop an obsession with water and lurk around faucets or water bowls. Appetite is suddenly either ravenous (up to three-times normal) or absent. These symptoms arise from the body being unable to use glucose as an energy source.

A fasting glucose blood test will normally be suggestive of diabetes at this point. The same home blood test monitors used in humans are used on cats, usually by obtaining blood from the ear edges or paw pads. As the disease progresses, ketone bodies will be present in the urine, which can be detected with the same urine strips as in humans.

In the final stages, the cat starts wasting and the body will breaking down its own fat and muscle to survive. Lethargy or limpness, and acetone-smelling breath are acute symptoms of ketoacidosis and dehydration and is a medical emergency.

Untreated, diabetes leads to coma and then death.

Diabetic Emergencies

Too little insulin over time can cause tissue starvation (as glucose can't reach the brain or body). In combination with dehydration, fasting, infection, or other body stresses, this can turn over a few hours into diabetic ketoacidosis, a medical emergency with a high fatality rate, that cannot be treated at home. Many undiagnosed diabetic cats first come to the vet in this state, since they haven't been receiving insulin. Symptoms include lethargy, acetone or fruity smell on breath, shortness of breath, high blood sugar, huge thirst drive. Emergency care includes fluid therapy, insulin, management of presenting symptoms and 24-hour hospitalization.

Complications

The back legs may become weak and the gait may become stilted or wobbly, due to diabetic neuropathy, which is caused by damage to the myelin sheath of the peripheral nerves due to glucose toxicity and cell starvation, which are in turn caused by chronic hyperglycemia. Most common in cats, the back legs become weaker until the cat displays a plantigrade stance, standing on its hocks instead of on its toes as normal. The cat may also have trouble walking and jumping, and may need to sit down after a few steps. Neuropathy sometimes heals on its own within 6–10 weeks once blood sugar is regulated.

Causes

The signs of diabetes mellitus are caused by a persistently high blood glucose concentration, which may be caused by either insufficient insulin, or by a lack of response to insulin. Most cats have a type of diabetes mellitus similar to human diabetes mellitus type 2, with β-cell dysfunction and insulin resistance. Factors which contribute to insulin resistance include obesity and endocrine diseases such as acromegaly. Acromegaly affects 20–30% of diabetic cats; it can be diagnosed by measuring the concentration of insulin-like growth factor-1 (IGF-1) in the blood.

Management

Diabetes can be treated but is life-threatening if left alone. Early diagnosis and treatment by a qualified veterinarian can help in preventing nerve damage, and, in rare cases, lead to remission. Cats do best with long-lasting insulin and low carbohydrate diets. Because diabetes is a disease of carbohydrate metabolism, a move to a primarily protein and fat diet reduces the occurrence of hyperglycemia.

Diet

Diet is a critical component of treatment, and is in many cases effective on its own. For example, a recent mini-study showed that many diabetic cats stopped needing insulin after changing to a low carbohydrate diet. The rationale is that a low-carbohydrate diet reduces the amount of insulin needed and keeps the variation in blood sugar low and easier to predict. Also, fats and proteins are metabolized slower than carbohydrates, reducing dangerous blood-sugar peaks right after meals.

Recent recommended diets are trending towards a low carbohydrate diet for cats rather than the formerly-recommended high-fiber diet. Carbohydrate levels are highest in dry cat foods made out of grains (even the expensive "prescription" types) so cats are better off with a canned diet that is protein and fat focused. Both prescription canned foods made for diabetic cats and regular brand foods are effective. Owners should aim to supply no more than 10% of the daily energy requirement of cats with carbohydrates.

Medications

Oral medications like Glipizide that stimulate the pancreas, promoting insulin release (or in some cases, reduce glucose production), are less and less used in cats, and these drugs may be completely ineffective if the pancreas is not working. These drugs have also been shown in some studies to damage the pancreas further or to cause liver damage. Some owners are reluctant to switch from pills to insulin injections, but the fear is unjustified; the difference in cost and convenience is minor (most cats are easier to inject than to pill), and injections are more effective at treating the disease.

Insulin

The method usually employed is a dose of slow-acting insulin, twice daily, to keep the blood sugar within a recommended range for the entire day. With this method, it is important for the cat to avoid large meals or high-carbohydrate food. Meals may also be timed to coincide with peak insulin activity. Once-daily doses are not recommended, since insulin usually metabolizes faster in cats than in humans or dogs. For example, an insulin brand that lasts 24 hours in people may only be effective for about 12 in a cat.

Cats may be treated with animal insulin (bovine-based insulin is most similar to cat insulin), or with human synthetic insulin. The best choice of insulin brand and type varies from animal-to-animal and may require some trial-and-error. The human synthetic insulin, Humulin N /Novolin N/ NPH, is usually a poor choice for cats, since cats metabolize insulin about twice as fast. The Lente and Ultralente versions were popular for feline use until summer 2005, when they were discontinued.

Until the early 1990s, the most recommended type for pets was bovine/porcine-derived PZI, but that type was phased out over the 1990s and is now difficult to find in many countries. There are sources in the US and UK, and many vets are now starting to recommend them again for pets, but they have been discontinued by most manufacturers as of 2007-2008. A new synthetic PZI analogue called ProZinc is now available.

Caninsulin (known in the US as Vetsulin) is a brand of porcine-based insulin approved for cats which is available with a veterinarian's prescription. According to the manufacturer's website, the insulin's action profile in cats was similar to that of NPH insulin, and lowered blood sugar quickly, but for only about 6–8 hours. Vetsulin was recalled in the US in November 2009 due to inconsistent strength; it was available again as of April 2013.

Two ultra-slow time-release synthetic human insulins became available in 2004 and 2005, generically known as insulin detemir (Levemir) and insulin glargine (Lantus). Studies have had good results with insulin glargine in cats. Follow-up research shows that Levemir can be used with a similar protocol and that either insulin, on this protocol, can lead uncomplicated feline cases to remission, with the most success being in cats who start on these protocols as soon as possible after diagnosis.

Dosage and Regulation

Cats may have their mealtimes strictly scheduled and planned to match with injection times, especially when on insulin with a pronounced peak action like Caninsulin/Vetsulin or Humulin N. If the cat free-feeds and normally eats little bits all day or night, it may be best to use a very slow-acting insulin to keep a constant level of blood glucose. Some veterinarians still use the outdated recommendation of using Humulin "N" or NPH insulin for cats, which is very fast-acting for most cats. The slower-acting Lente and Ultralente (Humulin L and Humulin U) insulins were discontinued in 2005 so most cats are treated with either the veterinary PZI insulins, or the new full-day analogs glargine (Lantus) and detemir (Levemir).

The first goal is to regulate the cat's blood glucose by keeping the blood glucose values in a comfortable range for the cat during the most of the day. This may take a few weeks to achieve.

The most successful documented method is tight regulation with Lantus or Levemir.

Typical obstacles to regulation include:

- Chronic overdose masked by Somogyi: A dose that is too high may cause a Somogyi rebound, which can look like a need for more insulin. This condition can continue for days or weeks.

- High-carbohydrate cat food: Many commercial foods (especially light foods) are very high in carbohydrates. The extra carbohydrates keep the cat's blood sugar high. In general, canned foods are lower in carbohydrates than dry foods, and canned "kitten" foods lower still. Diabetes in cats can be better regulated and even sometimes reversed with a low carbohydrate diet.

- Inappropriate insulin: Different brands and types of insulin have idiosyncratic effects on different cats. With some dosages, the insulin may not last long enough for the cat. Testing blood sugar more frequently can determine if the insulin is controlling the blood sugar concentration throughout the day.

Blood Sugar Guidelines

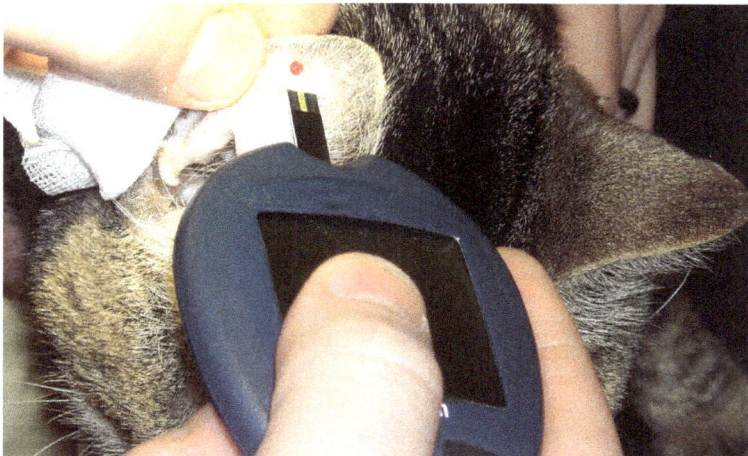

Taking a blood sample from a cat's ear to measure blood glucose concentration on a glucometer.

Absolute numbers vary between pets, and with meter calibrations. Glucometers made for humans are generally accurate using feline blood except when reading lower ranges of blood glucose (<80 mg/dl–4.44 mmol/L). At this point the size difference in human and animal red blood cells can create inaccurate readings.

Somogyi Rebound

Too much insulin may result in a contradictory increase of blood glucose. This "Somogyi effect" is often noted by cat owners who monitor their cat's blood glucose at home. Anytime the blood glucose level drops too far to hypoglycemia, the body may defensively dump glucose (converted from glycogen in the liver), as well as hormones epinephrine and cortisol, into the bloodstream. The glycogen raises the blood glucose, while the other hormones may make the cat insulin-resistant for a time. If the body has no glycogen reserves, there will be no rebound effect and the cat will just be hypoglycemic.

Even a small overdose can trigger a rebound effect (A typical case is increasing bidaily dosage from 1 unit to 2, passing a correct dose of 1.5 units.)

Rebound hyperglycemia occurs rarely in cats treated with glargine in a protocol aiming for tight control of blood glucose concentrations.

Hypoglycemia

An acute hypoglycemic episode (very low blood sugar) can happen to even careful pet owners, since cats' insulin requirements sometimes change without warning. The symptoms include depression/lethargy, confusion/dizziness, loss of excretory/bladder control, vomiting, and then loss of consciousness and seizures. Immediate treatment includes administering honey or corn syrup by rubbing on the gums of the cat (even if unconscious, but not if in seizures). Symptomatic hypoglycemia in cats is a medical emergency and the cat will require professional medical attention. The honey/corn syrup should continue to be administered on the way to the vet, as every minute without blood sugar causes brain damage.

A cat with hypoglycemia according to a blood glucose meter (<2.2 mmol/L or 40 mg/dL), but with no symptoms, should be fed as soon as possible. Hypoglycemic cats that refuse to eat can be force-fed honey or corn syrup until they stabilize.

Mild hypoglycemic episodes can go unnoticed, or leave evidence such as urine pools outside the litter box. In these cases the blood sugar will probably appear paradoxically high upon the next test hours later, since the cat's body will react to the low blood sugar by stimulating the liver to release stored glycogen.

Remission

Remission occurs when a cat no longer requires treatment for diabetes mellitus, and has normal blood glucose concentrations for at least a month.

Approximately one in four cats with type 2-like diabetes achieve remission. Some studies have reported a higher remission rate than this, which may in part be due to intensive monitoring that is impractical outside of a research environment. Research studies have implicated a variety of factors in successful remission; in general, the following factors increase the likelihood of remission:

- Diabetes was diagnosed a few months ago.

- The cat has no other serious disease.

- Treatment includes insulin glargine administered twice daily.

- The cat is monitored frequently during the first few months of treatment.

- The cat eats a diet low in carbohydrates and high in protein.

Cats may present with type-2 (insulin-resistant) diabetes, at least at first, but hyperglycemia and amyloidosis, left untreated, will damage the pancreas over time and progress to insulin-dependent diabetes.

Glipizide and similar oral diabetic medicines designed for type-2 diabetic humans have been shown to increase amyloid production and amyloidosis, and therefore may reduce likelihood of remission.

Approximately one third of cats which achieve remission will later relapse.

Feline Asthma

Feline asthma is a common allergic respiratory disease in cats, affecting at least one percent of all adult cats worldwide. It is a chronic progressive disease for which there is no cure. Common symptoms include wheezing, coughing, labored breathing and potentially life-threatening bronchoconstriction. There is conjecture that the disease is becoming more common due to increased exposure to industrial pollutants.

Signs and Symptoms

Feline asthma occurs with the inflammation of the small passageways of a cat's lungs, during the attack the lungs will thicken and constrict making it difficult to breathe. Mucus may be released by the lungs into the airway resulting in fits of coughing and wheezing. Some cats experience a less severe version of an asthma attack and only endure some slight coughing. The obvious signs that a cat is having a respiratory attack are: coughing, wheezing, blue lips and gums, squatting with shoulders hunched and neck extended, rapid open mouth breathing or gasping for air, gagging up foamy mucus and overall weakness.

Diagnosis

Owners often notice their cat coughing several times per day. Cat coughing sounds different from human coughing, usually sounding more like the cat is passing a hairball. Veterinarians will classify the severity of feline asthma based on the medical signs. There are a number of diseases that are very closely related to feline asthma which must be ruled out before asthma can be diagnosed. Lungworms, heartworms, upper and lower respiratory infections, lung cancer, cardiomyopathy and lymphocytic plasmacytic stomatitis all mimic asthmatic symptoms. Medical signs, pulmonary radiographs, and a positive response to steroids help confirm the diagnosis.

While radiographs can be helpful for diagnosis, airway sampling through transtracheal wash or bronchoalveolar lavage is often necessary. More recently, computed tomography has been found to be more readily available and accurate in distinguishing feline tracheobronchitis from bronchopneumonia.

Treatment

Although feline asthma is incurable, ongoing treatments allow many domestic cats to live normal lives. Feline asthma is commonly managed through use of bronchodilators for mild cases, or glucocorticosteroids with bronchodilators for moderate to severe cases.

Previously, standard veterinary practice recommended injected and oral medications for control of the disease. These drugs may have systemic side effects including diabetes and pancreatitis. In 2000, Dr. Philip Padrid pioneered inhaled medications using a pediatric chamber and mask using

Flovent (fluticasone) and salbutamol. Inhaled treatments reduce or eliminate systemic effects. In 2003 a chamber called the AeroKat Feline Aerosol Chamber was designed specifically for cats, significantly improving efficiency and reducing cost for the caregiver. Medicine can also be administered using a human baby spacer device. Inhaled steroid usually takes 10–14 days to reach an effective dose.

Prevention

Feline asthma and other respiratory diseases may be prevented by cat owners by eliminating as many allergens as possible. Allergens that can be found in a cat's habitual environment include: pollen, molds, dust from cat litter, perfumes, room fresheners, carpet deodorizers, hairspray, aerosol cleaners, cigarette smoke, and some foods. Avoid using cat litters that create lots of dust, scented cat litters or litter additives. Of course eliminating all of these can be very difficult and unnecessary, especially since a cat is only affected by one or two. It can be very challenging to find the allergen that is creating asthmatic symptoms in a particular cat and requires a lot of work on both the owner's and the veterinarian's part. But just like any disease, the severity of an asthma attack can be propelled by more than just the allergens, common factors include: obesity, stress, parasites and pre-existing heart conditions. Dry air encourages asthma attacks so keep a good humidifier going especially during winter months.

Cat Skin Disorders

Cat skin disorders are among the most common health problems in cats. Skin disorders in cats have many causes, and many of the common skin disorders that afflict people have a counterpart in cats. The condition of a cat's skin and coat can also be an important indicator of its general health. Skin disorders of cats vary from acute, self-limiting problems to chronic or long-lasting problems requiring life-time treatment. Cat skin disorders may be grouped into categories according to the causes.

Types of Disorders

Immune-mediated Skin Disorders

Skin disease may result from deficiencies in immune system function. In cats, the most common cause of immune deficiency is infection with retroviruses, FIV or FeLV, and cats with these chronic infections are subject to repeated bouts of skin infection and abscesses. This category also includes hypersensitivity disorders and eosinophilic skin diseases such as atopic dermatitis, miliary dermatitis and feline eosinophilic granuloma and skin diseases caused by autoimmunity, such as pemphigus and discoid lupus.

Infectious Skin Diseases

An important infectious skin disease of cats is ringworm, or dermatophytosis. Other cat skin infections include parasitic diseases like mange and lice infestations.

Other ectoparasites, including fleas and ticks, are not considered directly contagious but are acquired from an environment where other infested hosts have established the parasite's life cycle.

Another common skin infection is cat bite abscess. A mixture of bacteria introduced by a bite wound cause infections in pockets under the skin and affected cats often show manic depression and fever.

Hereditary and Developmental Skin Diseases

Some diseases are inherent abnormalities of skin structure or function. These include skin fragility syndrome (Ehlers-Danlos), hereditary hypotrichosis and congenital or hereditary alopecia.

Cutaneous Manifestations of Internal Diseases

Some systemic diseases can become symptomatic as a skin disorder. In cats this includes one of the most devastating cat skin disorders, feline acquired skin fragility syndrome, which can come from starvation or over-treatment with cortisone-like drugs or with diabetes, FIP or Cushing's Disease.

Nutrition Related Disorders

Nutritional related disorders can arise if the cat's food intake decreases, interactions between ingredients or nutrients occur, or mistakes are made during food formulation or manufacturing. Degradation of some nutrients can occur during storage. Nutritional related skin disorders can result in excesses or deficiencies in the production of sebum and in keratinization, the toughening of the outer layer of the skin. This can result in dandruff, erythema, hair loss, greasy skin, and diminished hair growth.

Minerals

Zinc is important for the skin's function, as it is involved in the production of DNA and RNA, and therefore important for cells that divide rapidly. A deficiency in zinc mainly results in skin disorders in adult cats, but also results in growth oddities. The skin of a cat deficient in zinc would likely have erythema and hair loss. The cat may have crusty, scaly skin on its limbs or tail. The coat of the cat becomes dull. Similarly, copper can affect coat health of cats; deficiencies will cause fading of coat color and weakened skin, leading to lesions.

Protein

The hair of a cat is made of mainly protein, and cats need about 25-30% protein in their diets, much higher than what a dog needs. A deficiency in protein usually happens when kittens are fed dog food or when low-protein diets are fed improperly. If a cat has a protein deficiency, the cat will lose weight. The coat condition will be poor, with dull, thinning, weak, and patchy hair. To remedy this, a diet with adequate amounts of protein must be fed.

Essential Fatty Acids

Cats must have both linoleic acid and arachidonic acid in their diet, due to their low production of the δ-6 desaturase enzyme. A deficiency in these fatty acids can occur if the fats in the cat's food are oxidized and become rancid from improper storage. A cat will be deficient for many months prior to seeing clinical signs in the skin, after which the skin will become scaly and greasy, while the coat will become dull. To treat health concerns caused by a deficiency of fatty acids, the ratio of n-3 to n-6 fatty acids must be corrected and supplemented.

Vitamin A

Cats cannot synthesize vitamin A from plant beta-carotene, and therefore must be supplemented with retinol from meat. A deficiency in vitamin A will result in a poor coat, hair loss, and scaly, thickened skin. However, an excess of vitamin A, called hypervitaminosis A, can result from over feeding cod liver oil and large amounts of liver. Signs of hypervitaminosis A are overly sensitive skin and neck pain, causing the cat to be unwilling to groom itself, resulting in a poor coat. Supplementing vitamin A with retinol to a deficient cat and feeding a balanced diet to a cat with hypervitaminosis A will treat the underlying nutritional disorder.

Vitamin B

The cat must have a supply of niacin, as cats cannot convert tryptophan into niacin. However, diets high in corn and low in protein can result in skin lesions and scaly, dry, greasy skin with hair loss. A deficiency of the B vitamin biotin causes hair loss around the eyes and face. A lack of B vitamins can be corrected by supplementing with a vitamin B complex and brewer's yeast.

Cat Flu

Cat flu, or upper respiratory infection (URI) is a very common disease that can vary considerably in severity, and on occasions can even be life-threatening.

In the vast majority of cases, disease results from infection with feline calicivirus (FCV) or feline herpes virus (FHV, or FHV-1). Clinical signs include sneezing, nasal discharge, conjunctivitis (inflammation of the lining of the eyes), ocular discharge, loss of appetite, fever and depression. Mouth ulcers, coughing, excessive drooling of saliva and eye ulcers may also be seen. Very young, very old and immunosuppressed cats are more likely to develop severe disease and possibly die as a result of their URI, usually due to secondary infections (such as pneumonia), lack of nutrition and dehydration.

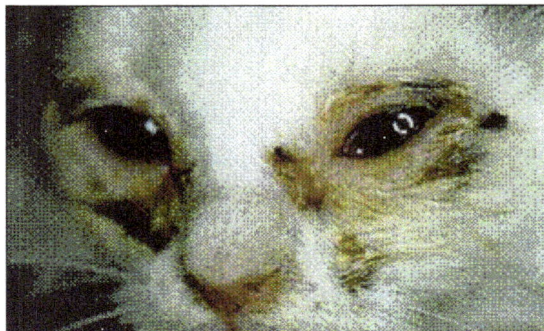

Typical ocular and nasal discharges of cat flu.

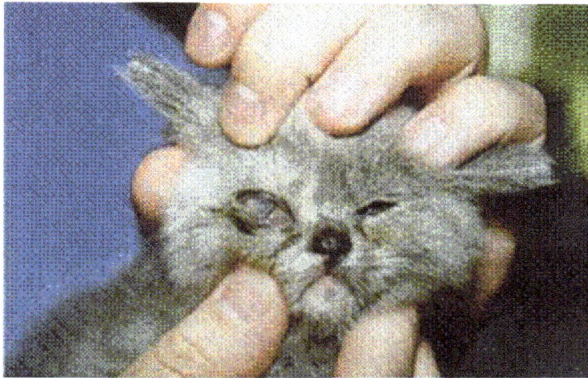

Infection with feline herpes virus can cause serious eye damage.

What Cats are at Risk of Uris?

URIs are common, as the causative viruses are widespread in cat populations. Typical risk factors include:

- Cats kept in large groups or colonies such as breeding catteries, rescue centres and feral cat colonies in these situations the viruses are able to spread easily.

- Unvaccinated cats.

- Kittens.

- Elderly and immunosuppressed cats (e.g., cats with FeLV or FIV infection, or cats receiving immunosuppressive therapy) are more vulnerable to developing severe disease.

Causes of URIs in Cats

Most cat URIs are caused by infection with one or both of the cat flu viruses:

- Feline herpes virus (FHV or FHV-1, formerly known as feline rhinotracheitis virus).

- Feline calicivirus (FCV).

These two viruses are thought to be responsible for more than 90% of URIs in cats. Other important organisms that may be involved in some cases include:

- *Bordetella bronchiseptica* (may be a cause of sneezing, nasal discharge and sometimes coughing important in some colony situations).

- *Chlamydophila felis* (this is mainly a cause of ocular disease conjunctivitis).

Clinical Signs of Uri

The incubation period following infection with FCV or FHV is usually just a few days (2-10 days). After this, typical clinical signs develop which include:

- Sneezing.

- Nasal discharge.

- Ocular discharge.

- Lethargy.

- Inappetence.

- Fever.

The severity of these signs varies considerably in some cats the signs are very mild and transient, in others they may be very marked and severe. There are some differences in clinical presentation between the two viruses, but these are not sufficient to be able to distinguish them simply from clinical signs:

- FHV infection tends to be more severe, often causing more marked conjunctivitis (eye infection and ocular discharge), and some ulceration of the cornea (the clear part at the front of the eye). FHV may also cause: severe pharyngitis leading to anorexia; inflammation in the trachea; and coughing.

- FCV infection is often milder, with inapparent or less severe ocular signs, but FCV often causes ulceration of the tongue (and sometimes the palate or the lips). FCV may cause a transient arthritis (limping syndrome), usually seen in young kittens, and in very young kittens can cause severe viral pneumonia.

Although FCV and FHV are viral infections, secondary infection with bacteria is common and can contribute to rhinitis (infection in the nose) conjunctivitis, and even lung infections. While most cats will recover from URIs, on occasions they can be life-threatening, and with severe infections the recovery may take several weeks. Some cats may also be left with permanent damage within the nose and may have persistent or recurrent nasal discharge (so-called 'chronic rhinitis').

In rare cases, a much more severe and often fatal form of FCV infection may occur. This is associated with particular strains of the virus that are highly virulent and termed 'virulent systemic FCV' (vsFCV) infection. Fortunately such infections are very rare.

Diagnosis

Diagnosis by your vet is usually based on the typical signs associated with URIs, and exclusion of other causes. It is possible to confirm a diagnosis and to investigate which viruses are involved, but this is often not necessary.

Testing for FCV or FHV involves collecting a mouth or eye swab which is then sent to a specialised veterinary laboratory. Here the virus can be identified through culturing or by a PCR test (a molecular test to show the presence of the viral genes).

Treatment

Treatment of URIs is largely symptomatic and supportive. Your vet may want to do some additional tests if they are concerned about the extent of disease (e.g., the possibility of pneumonia) or if they are worried about complications (such as infection with FIV or FeLV).

Antibiotics are indicated to treat secondary bacterial infections and to try to reduce the damage the infection causes. If nasal congestion is severe and breathing is difficult your vet may also suggest steam inhalation or nebulisation make discharges more liquid and more easily relieved by sneezing.

Affected cats are often reluctant to eat they will have a poor sense of smell and eating may also be uncomfortable. Using soft, highly aromatic foods (for example kitten foods, fish in oil) that are gently warmed will help to tempt an inappetent cat. However, if anorexia is severe your cat may require hospitalisation for your vet to provide food via a feeding tube. This can be important, as poor nutrition will significantly contribute to disease and slow down healing. Intravenous fluids may also be needed if your cat is not drinking properly, to avoid dehydration. Analgesics may also be required.

Interferons are proteins that are produced in the body, in part to help fight viral infections. Injectable interferons may be used as a supportive treatment (either high doses of recombinant human interferons or recombinant feline interferon) there is some evidence that this may be of benefit, but it probably needs to be given early in the course of disease for best effect.

There are a number of topical antiviral agents that can help to manage FHV-associated ocular disease (such as trifluoridine, idoxuridine and cidofovir). More recently a drug used to treat human herpes virus infections famciclovir – has been shown to be safe and effective in cats when given orally. This is a major step forward in managing severe FHV infection in cats.

General nursing is also essential discharges from the eyes and nose should be gently wiped away using damp cotton wool, and the cat should be kept warm and comfortable.

Carriers

Most cats that recover from infection with URI viruses will become 'carriers'. Carrier cats usually show no sign of illness but, may shed virus in saliva, tears and nasal secretions, and can be a source of infection to other cats.

Although almost all cats infected with FHV will remain long-term carriers, many of these will never shed significant amounts of virus. Others may shed virus intermittently, especially during times of stress. Some cats may show mild signs of URI again when they shed the virus, but most do not. Carrier cats in a breeding colony are a source of risk to their kittens, as the stress of kittening may induce shedding of FHV.

Most cats infected with FCV remain carriers of the virus, and continue to shed the virus for a period of weeks or months after infection, but the majority (although not all) will eventually eliminate the virus within a few months.

Spread of Infection

The viruses associated with URIs are spread in three ways:

- Direct contact with an infected cat showing signs of URI.

- Direct contact with a carrier cat shedding virus.

- Contact with virus carried on clothing, food bowls and other objects. Large amounts of virus are present in the saliva, tears and nasal discharges and the viruses may be able to survive in the environment and on objects for up to 2 (FHV) to 10 (FCV) days.

Prevention

Vaccination: The risk of URIs can be dramatically reduced by vaccination against FHV and FCV. These vaccines are important for all cats, irrespective of how they are kept (even if kept totally indoors), as the diseases are so ubiquitous. Although vaccination usually prevents severe disease developing, it cannot always prevent infection occurring and so mild disease may still develop in some cats. FCV has many different strains, and this can cause further problems with vaccination as vaccines will not necessarily work against all these strains. Newer FCV vaccines contain more than one strain in the vaccine to help overcome this problem. This is not an issue with FHV as only one virus strain exists.

Barrier nursing and disinfection: If there is more than one cat in a household, it is important to try to minimise the risk of infection being spread to the other cats. This is not always possible, but in addition to ensuring that all cats are vaccinated, where possible a cat showing clinical signs should be kept isolated from the other cats (e.g., confined to one room). Separate food bowls and litter trays should be used, and ideally the cat should be kept in a room that has very easy to disinfect surfaces (i.e., not soft furnishing and carpet). These viruses are susceptible to most disinfectants but make sure you check with your vet some disinfectants (such as phenolic-based products) are not safe to use around cats. Hypochlorite (bleach-based) disinfectants (eg, 5% bleach diluted 1:32) are effective against these viruses, but take care to use any disinfectant carefully most are irritant to cats if they come into direct contact with the disinfectant.

References

- "Fibrosing Cardiomyopathy /Center for Academic Research and Training in Anthropogeny (CARTA)". Carta. anthropogeny.org. Center for Academic Research and Training in Anthropogeny. Retrieved 2019-05-15

- Non-human-primates, zoonoses, environmental-epidemiology: vdh.virginia.gov, Retrieved 9 May, 2019

- Rush, E. M., Ogburn, A. L., & Monroe, D. (2011). Clinical management of a western lowland gorilla (Gorilla gorilla gorilla) with a cardiac resynchronization therapy device. Journal of Zoo and Wildlife Medicine, 42(2), 263-276

- A-guide-to-common-horse-diseases: alansfactoryoutlet.com, Retrieved 10 July, 2019

- Macgillivray, Katherine Cole; Raymond W. Sweeney; Fabio Del Piero (July 2002). "Metastatic Melanoma in Horses". Journal of Veterinary Internal Medicine. 16 (4): 452–456. Doi:10.1111/j.1939-1676.2002.tb01264.x

- Cat-flu-upper-respiratory-infection, advice: icatcare.org, Retrieved 11 August, 2019

- Chambers, G; V. A. Ellsmore; P. M. O'Brien; S. W. J. Reid; S. Love; M. S. Campo; L. Nasir (May 2003). "Association of bovine papillomavirus with the equine sarcoid". Journal of General Virology. 84 (5): 1055–1062. Doi:10.1099/vir.0.18947-0. PMID 12692268. Retrieved 6 August 2011

- Chrisman, Cheryl; Clemmons, Roger; Mariani, Christopher; Platt, Simon (2003). Neurology for the Small Animal Practitioner (1st ed.). Teton New Media. ISBN 1-893441-82-2

PERMISSIONS

We would like to thank the editorial team for lending their expertise to make the book truly unique. They have played a crucial role in the development of this book. Without their invaluable contributions this book wouldn't have been possible. They have made vital efforts to compile up to date information on the varied aspects of this subject to make this book a valuable addition to the collection of many professionals and students.

This book was conceptualized with the vision of imparting up-to-date and integrated information in this field. To ensure the same, a matchless editorial board was set up. Every individual on the board went through rigorous rounds of assessment to prove their worth. After which they invested a large part of their time researching and compiling the most relevant data for our readers.

The editorial board has been involved in producing this book since its inception. They have spent rigorous hours researching and exploring the diverse topics which have resulted in the successful publishing of this book. They have passed on their knowledge of decades through this book. To expedite this challenging task, the publisher supported the team at every step. A small team of assistant editors was also appointed to further simplify the editing procedure and attain best results for the readers.

Apart from the editorial board, the designing team has also invested a significant amount of their time in understanding the subject and creating the most relevant covers. They scrutinized every image to scout for the most suitable representation of the subject and create an appropriate cover for the book.

The publishing team has been an ardent support to the editorial, designing and production team. Their endless efforts to recruit the best for this project, has resulted in the accomplishment of this book. They are a veteran in the field of academics and their pool of knowledge is as vast as their experience in printing. Their expertise and guidance has proved useful at every step. Their uncompromising quality standards have made this book an exceptional effort. Their encouragement from time to time has been an inspiration for everyone.

The publisher and the editorial board hope that this book will prove to be a valuable piece of knowledge for students, practitioners and scholars across the globe.

INDEX

www.ingramcontent.com/pod-product-compliance
Lightning Source LLC
Chambersburg PA
CBHW061255190326
41458CB00011B/3673